About the author

James Smith is professor of African and development studies in the Centre of African Studies. He is also a director at the ESRC Innogen Research Centre at Edinburgh and a visiting fellow in development policy and practice at the Open University. His research explores the relationships between knowledge, science and development, particularly in relation to agricultural research and how it is practised. He has worked with many international organizations and research centres, including Oxfam, DfID, IDRC and the Consultative Group on International Agricultural Research.

Biofuels and the globalization of risk

the biggest change in North–South relationships since colonialism?

James Smith

Zed Books
LONDON · NEW YORK

Biofuels and the globalization of risk: the biggest change in North–South relationships since colonialism? was first published in 2010 by Zed Books Ltd, 7 Cynthia Street, London N1 9JF, UK and Room 400, 175 Fifth Avenue, New York, NY 10010, USA

www.zedbooks.co.uk

Set in Monotype Sabon and Gill Sans Heavy by Ewan Smith, London
Index: ed.emery@thefreeuniversity.net
Cover designed by Rogue Four Design

A catalogue record for this book is available from the British Library
Library of Congress Cataloging in Publication Data available

ISBN 978 1 84813 571 0 hb
ISBN 978 1 84813 572 7 pb
ISBN 978 1 84813 573 4 eb

Contents

Acknowledgements

The research and the time necessary to write this book have been primarily supported by two grants. I would like to acknowledge the UK Economic and Social Research Council and their grant to the ESRC Innogen Centre. I would also like to acknowledge the UK Department for International Development for funding our five-year PISCES project (Policy Innovation Systems for Clean Energy Sustainability), which has given me the opportunity to meet many new colleagues from many parts of the world. Thanks to my colleagues at the African Centre of Technology Studies, Kenya, the University of Dar es Salaam, Tanzania, the M. S. Swaminathan Research Foundation, India, and Practical Action Consulting, UK and Sri Lanka. All opinions and errors are, of course, mine.

I would like to thank several of my colleagues for wisdom and assistance: Norman Clark, Lawrence Dritsas, Steven Hunt, Tom Molony, Arivudai Nambi, Colin Pritchard and Judi Wakhungu. In particular I would like to thank some of my (current and ex-) doctoral students who have helped in many ways, in particular Maija Hirvonen, Shishusri Pradhan and Shaun Ruysenaar.

Finally, I would like to dedicate this book to my partner, Barbara, whose interest in liquid biofuels extends about as far as a good South African Merlot, but whose interest in social justice, international development and Africa extends to all horizons.

Acronyms

CGIAR	Consultative Group on International Agricultural Research
CIAT	International Center for Tropical Agriculture
CIFOR	Center for International Forestry Research
CIMMYT	International Maize and Wheat Improvement Center
ECOWAS	Economic Community of West African States
EPA	Environmental Protection Agency
EROI	energy return on investment
EU	European Union
FAO	Food and Agriculture Organization
FDI	foreign direct investment
GHG	greenhouse gas
GM	genetically modified
ICRISAT	International Crops Research Institute for the Semi-Arid Tropics
LCA	life-cycle analysis
NGO	non-governmental organization
OECD	Organisation for Economic Co-operation and Development
R&D	research and development
RTFO	Renewable Transport Fuel Obligations
UK	United Kingdom
UN	United Nations
UNCTAD	United Nations Conference on Trade and Development
UNDP	United Nations Development Programme
UNEP	United Nations Environment Programme
USA	United States of America
WTO	World Trade Organization

ONE
Introduction: perfect storms

In 2009, John Beddington, the UK Government Chief Scientific Adviser, adopted the term 'perfect storm' to describe a future global confluence of food, water and energy insecurity (2009). While Beddington captured something of the unprecedented nature and threat of the future, the choice of 'storm' as a sort of metaphorical de-anthropomorphism ran somewhat counter to the usual humanizing of extreme weather events; for we need to be absolutely clear that the processes and interactions that will lead to food, water and energy insecurity are primarily driven by us, our exigencies and the choices we make. A great many of these interactions hinge on our trajectories of development and wielding of new knowledge and new technologies.

It is becoming increasingly apparent, and increasingly globally apparent, that we must learn how to better harness science and technology to these ends, and accordingly we need to understand the complex, interrelated contexts and processes that lead to new technologies, new priorities and new directions that will serve our future and deal with, rather than exacerbate, present and future risks. Biofuels − liquid fuels that are directly derived from renewable biological resources and especially from purpose-grown crops − throw many of these issues into the sharpest relief. One of the most striking features of biofuels as a global solution is their huge potential to entirely reshape livelihoods, patterns of resource consumption, environments and agro-food production systems; there is a cost for every benefit, and that is often invisible under the veneer of technological promise.

Much as the technological optimism of biofuels shifts responsibility towards the immediate futures of others, biofuels risk exporting impact and risk elsewhere. The production of biofuels risks reprioritizing land use across the globe, and as yet we know relatively little about the implications of this. Biofuels are driving, and transforming, the increasingly entangled relationship between energy, food security and climate change, and consequently trying to understand the politics, narratives and discourses that drive policy and practice surrounding biofuels provides an opportunity to reflect on the thorny relationship between science, development and the environment (Molony and Smith 2010).

In some respects biofuels are simple technologies. We are simply deriving energy from plants, through seed oil or biomass, primarily to combust in car engines. Future, better technologies may unlock efficiencies or new ways of deriving energy from plants, but the basic principle remains. In other respects biofuels are extremely complex. They are being developed within complex systems and their production in itself creates new complex systems. They generate couplings between agricultural systems, international markets, petrochemical companies, consumers and producers. These couplings have implications of their own. Charles Perrow, in his book *Normal Accidents* (1999), analyses the implications of the unexpected consequences of technological interactions taking place within what he terms 'tightly coupled' systems. These systems are so interrelated and complex that there is no easy way to control or contain negative consequences once a technology has started to unravel. This unravelling may quickly become irretrievable, and indeed any attempts to intervene may exacerbate the problem if we simply do not sufficiently understand the mechanics or truly recognize the root of the problem. The lure of biofuels may blind us to the risks bound up in the intricacies of new technologies and the limits of our ability to deal with them. We may understand the chemistry of photosynthesis and the physics of combustion, but we may not understand the gamut of interactions and implications necessary to efficiently join them up.

Biofuels represent both a promise of a technologically driven future and the spectre of a Rumsfeldian web of known unknowns and unknown unknowns (and presumed knowns). They represent an increasingly globalized, interconnected world, and a world where production, risk and responsibility are inherently localized, constantly diverted and increasingly entangled. Biofuels represent at the same time an effort to acknowledge and deal with some of the most pressing global problems we face, and an excuse to not really deal directly with these problems, or even to really *understand* them, and in particular what drives them. Biofuels, their development, their deployment, the ideas they represent and the sorts of solutions they suggest, are rooted in the contradictory processes of global progress, consumption and development. They reflect how we imagine the world to be, and refract how we – or those who make decisions at least – imagine it ought to be, or can be. This book will focus on the recent development of biofuels as both a solution to, and a driver of, perfect storms.

Biofuelled futures

In the space of a few years biofuels have shifted from existing beneath the radar of development, to being seen as a possible multi-purpose solution to a range of problems – climate change, energy insecurity and rural underdevelopment – to representing a 'crime against humanity', according to the UN Special Rapporteur on the Right to Food, owing primarily to the perceived impact of investment in biofuel production on food stocks and subsequently on global food prices.[1] These contested and shifting perceptions have done little to substantially decelerate biofuels as a policy idea or as an investment opportunity. Figures from the US Department of Agriculture for 2009 show that the grain grown to produce fuel was enough to feed 330 million people for one year at average consumption levels, according to the Earth Policy Institute.[2] This figure represents a third of all those who constantly go without sufficient food. In 2007, twenty-seven out of fifty countries surveyed either had policy under consideration or had enacted mandatory requirements for biofuels to be blended with traditional transport fuels, and forty had legislation to promote biofuels (Rothkopf 2007). Between 2002 and 2006, the amount of land used to grow biofuel crops quadrupled and production tripled (Coyle 2007).

David MacKay, in his excellent book *Sustainable Energy*, assesses the potential of biofuel production as a substitute for petrol in the UK (MacKay 2009). This is enlightening, if not particularly empowering. The average harvestable power of sunlight is 100 W/m^2

(watts per square metre). The most efficient plants in Europe are about 2 per cent efficient at turning solar energy into carbohydrates. This suggests that the most efficient plants might deliver 2 W/m^2, although in reality this translates into something nearer 0.5 W/m^2. MacKay assumes that if 75 per cent of the UK's land were devoted to bioenergy production, which would equate to 3,000m^2 of land planted with biofuel crops for every person, the energy output (ignoring any additional costs of growing, harvesting, processing and transporting) would be 36 kW (kilowatts) per day per person. The reality, if one were to attempt to assess the cumulative effect of inefficiencies throughout any processing chain, might – optimistically – be that an additional 33 per cent might be subtracted from that figure. To put this into context, a typical car-user uses about 40 kWh (kilowatt hours) per day; and if we fuel our car we still have to eat (ibid.).

In a similar vein, 'boundary-setting' calculations are compelling – and critical. Calculations indicate that all the vegetation in the USA contains only one third of the energy consumed in one year in the USA. That is *all* the vegetation, for current energy consumption patterns – not future demands – and does not account for any energy expended in the production, processing or transportation of biofuels (Pimentel, cited in Moore 2008). These statistics perhaps point to the potential of biofuels as hubris, or hot air.

These simple analyses encapsulate three important questions. First, why, if the potential for energy production (in the UK or USA at least) is so minuscule, are biofuels perceived to be so important? This is a key refrain of this book. Second, perhaps the UK's dismal weather means that biofuels can be produced more efficiently elsewhere? This is indeed true, and the 'comparative advantage' in terms of biomass production that the world's tropical regions enjoy has become a mobilizing factor in biofuels as a developmental tool.[3] Third, why, if the potential of bioenergy appears relatively limited, and risks trade-offs with other key determinants of human well-being, access to food primarily, is so much investment prompting turning agriculture from production of food to production of fuel? Good question.

In May 2008, the US Secretary of Agriculture claimed that their analysis showed that biofuel production contributed only 2–3 per cent to increases in food prices.[4] In July 2008, however, a World Bank document, dubbed the Mitchell Report, leaked to the *Guardian* newspaper, calculated that biofuel production was responsible for 75 per cent of the 140 per cent increase in staple food prices witnessed between 2002 and 2008 (Mitchell 2008). Increased biofuel production was said to have led to increased demand for so-called feedstock crops (from which fuel can be derived), which led to large-scale land-use changes at the expense of crops such as wheat for food consumption. The report concluded:

> The most important factor [in food price increases] was the large increase in biofuels production in the US and the EU. Without these increases, global wheat and maize stocks would not have declined appreciably, oilseed prices would not have tripled, and price increases due to other factors, such as droughts, would have been moderate. Recent export bans and speculative activities would probably not have occurred because they were largely responses to rising prices. (Ibid.)

Other organizations, such as Oxfam and the Organisation for Economic Co-operation and Development (OECD), estimated the impact of biofuel production on food price rises to be somewhere in between the incidental 3 per cent of the USA and the catastrophic 75 per cent of the Mitchell Report (Oxfam 2008). The variability of analysis is almost as striking as the attribution of impact. Why, when it is clear we simply do not know enough about the interactions, impacts and implications of massive investment in the production of biofuels, are we moving ahead so rapidly?

There are many reasons. Oil, the lubricant of the global economy, is running out. The concept of 'Peak Oil' has entered the popular lexicon, and conflict and instability in the major oil-producing regions of the world (notably the Middle East, Russia and Venezuela) have pushed the notion of energy security into the collective consciousness of policy-makers.[5] These concerns are allied to the rises in oil prices that we have witnessed over the past five or six years, which have only really been curbed by the global recession. We have more impetus than ever before to diversify our energy sources.

Additionally, and it has been a long time coming, an emergent global concern with the implications of climate change and a political consensus surrounding the role of humankind and particularly our dependence on fossil fuels have provided another powerful dimension for alternative energy sources.

A lack of access to energy is increasingly being framed as a fundamental impediment to development and consequently as a cornerstone of developing countries' poverty alleviation strategies: 'modern energy services are a powerful engine of economic and social development, no country has managed to develop much beyond a subsistence economy without ensuring at least minimum access to energy services for a broad section of its population' (FAO 2000: 1). Energy fuels development.

Alongside this, the production of biofuels and diversification into alternative energy sources more broadly are seen as strategic opportunities for rural areas to grow their economies, diversify their income sources and create jobs. This is a concern for both developed and developing countries. In OECD countries, the overproduction of agricultural commodities, low prices, under-utilized land, low farm incomes and powerful farm lobbies have created optimum conditions in which new agricultural commodity markets can be cultured. In the USA and the EU already heavy agricultural subsidies can be quite simply recalibrated towards the growing of crops for biofuel production. In developing countries, too, there are economic and developmental drivers of biofuel production, primarily reduced oil imports, rural development opportunities and subsequent opportunities for exports and income.

Biofuels fire the imagination of policy-makers, entrepreneurs, researchers and governments because of the possibility of being all things to all people. The corollary of this is, of course, that they risk becoming objects of contestation, or ideas around which ideologies and politics are fought, much as agricultural biotechnologies were before them. Arthur Mol (2007: 297) talks about the emergence of a global integrated biofuel network 'where environmental sustainabilities are more easily accommodated than vulnerabilities for marginal and peripheral groups and countries, irrespective of what policymakers and biofuels advocates tell us'. This global network – and its actors, aims, interests and ability to shape risk, impact and vulnerability in profound and complex new ways – lies at the heart of this book.

Globalizing technology and risk

A slew of recent research is grappling with how we can engender individual, institutional and international responsibility and action for climate change when we have collectively behaved so irresponsibly for so long (cf. Giddens 2009; Hulme 2009; Stern 2007). As recent history shows us, we have repeatedly and cumulatively struggled to come to terms with the implications of our economic and technological development, in terms of environmental impact, developmental outcome and social justice.

Truman's oft-cited, late-1940s presidential address emphasized the role of science and technology within development, and that relationship has been at the core of the global development project ever since: scientific advances would make available for all, for the first time, the tools necessary to relieve suffering and replace Hobbes with hope, and ignorance and tradition with knowledge and science. The 1950s and 1960s led to the Green Revolution and marked the interrelated globalization of development and an institutionalization of science and knowledge, which has profoundly shaped the relationship between science, technology and development up until the present day (Smith 2009).

In the last thirty years, however, science's ability to shackle the vagaries of the environment has been replaced with a postmodern, post-normal, existential view of new technologies as the progenitors of new types of risk (Beck 1992; Yearley 2005). An emerging critical view of science as no longer the sole arbitrator of fact or the producer of rationality is reflected in a world where values, perspectives and solutions are atomized and fragmented and where coming to terms with uncertainty through rationality seems increasingly counterintuitive (Beck 1992). Most recently, in early 2010, hacked email accounts, seemingly accident-prone (at best) peer review, and cumulative scientific error have cast climate scientists as people not to be fully trusted by the media. It might be altogether more productive and insightful to focus on the sheer complexity of climate systems and their interactions.

Rapid global change, driven by new economies and new technologies, fundamentally alters our relation to the world; risk is transported out of the sphere of things visited upon people through natural events and into the sphere of outcomes that are the result of often far-removed, collective decisions, or individual errors. We may no longer suffer famine in Europe, but neither did we risk enhanced global warming in the past. The immediate becomes prospective, and the proximate transforms into perspective. This is the conceptual context in which potentially transformative technologies such as biofuels must be viewed, and therefore understood.

Beddington's 'perfect storm' prompts us to think about a future where food insecurity and demands for energy and development have greatly increased, and in some senses competed, against a backdrop of increasing climatic and environmental uncertainty. In 2009, the Food and Agriculture Organization (FAO) reported that for the first time 1 billion people are undernourished, and this represents a significant recent increase to the steady-state figure of 'only' 800 million people that has been the norm for the past couple of decades (FAO 2009a). It is increasingly apparent that climate and food will become problems that affect us all, not simply a problem whose impacts will be confined to the global South. This realization of the globalization of risk into what Beck has termed the 'contract[ion] into a community of danger' (1992: 44) provides a powerful stimulus for change. Whether this will stimulate the sorts of profound political change that might lead to positive outcomes at climate change summits or undertakings to reorient agricultural subsidies away from production and protection remains to be seen; it is clear, however, that unknown risks, once perceived as sufficiently global, will stimulate immense new investment in science, in part as a direct solution and in part as an alternative to a political solution. This has important implications for biofuels, which concurrently represent both risk and response. The political economy of biofuels can drive forward entrenched and emerging sets of North–South and South–South relations (Dauvergne and Neville 2009). Amplifying existing power relations, or delineating new relationships based on historical inequities of power, risks entrenching risk in poorer countries and exacerbating historical patterns of resource consumption and exploitation.

The multidimensional potential of biofuels to contribute to energy security, sustainable development and climate change mitigation *without* forcing us to rethink existing patterns of global development, consumption and production is hugely alluring. In offering an ostensibly biotechnical solution to a highly political and deeply structural problem, the possibilities of biofuels offer a neat sidestep away from many of the knotty problems that underpin notions of sustainability or equity. It is clear from many of the so far little understood dynamics of food versus fuel debates that we risk becoming too attracted to the apolitics of biofuels and in doing so risk ignoring some of the more immediate problems, such as sufficiently understanding the nature of the complex biological, economic and political systems in which biofuels interact. We need policy that appreciates complexity and context, and recognizes the ways in which transformative technologies may transform in unexpected ways. We do not want, as Beck suggests we might, to risk replacing one sort of risk with another, particularly as this almost inevitably means transferring the burdens of risk to those least equipped to cope, as we have seen with the impacts of rises in global food prices. We may, unfortunately, be powerless to avoid it.

Thinking about risk is central to designing technologies that are appropriate and developing technologies that are sustainable. Doing this in the context of a complex, contested and rapidly changing world, driven by politics and interests, is not, however, straightforward. Thinking about the complex systems in which biofuels are developed, are promoted and impact upon us is crucial, and this involves thinking about new technologies and development in new, more nuanced ways.

Global assemblages

It is useful to think of the web of relationships in which biofuels exist in new ways, and to do this we need to step back and think about our collective relationship to technology. Bruno Latour has referred to science and technology as 'dark matter', because our lives are so firmly enmeshed with science and technology that we have spent relatively little energy in digesting the implications of this on our lives; we literally do not 'see' technology much of the time

(Latour 1992). It is only comparatively recently that scholars have begun to problematize science, its practice, knowledge, expertise, and its complex relationship and contribution to society. Alongside this, we have until very recently almost taken for granted the trajectories of globalization, the certainties of modernization and the pervasive spread of capitalism. These certainties coordinated our lives and our futures, shaped our responses to stimuli and change, mapped out our futures, and have perhaps contributed to our sense of (a lack of) responsibility for our actions.

We are beginning to think more carefully about globalization, progress and the future, though, and the seeming ubiquity of global change, and its drivers, has led to various critical analytic responses. One has been to develop meta-narratives of new orders, macro-processes or hegemonies, for example to think in terms of global dependency, world-systems theory or broad processes of globalization. A second response has been to focus on 'localities' as articulations with, hybridities of or sites of resistance to global forces (Escobar 2009). The Zapatistas of Chiapas state in Mexico are a notable example, but social or people's movements and the like are developing the world over. A third approach has sought to reconstitute the lens through which we understand the world, to focus on, for example, 'global' technology, culture or politics, increasingly adrift as they are from whatever roots and origins they may have once enjoyed (see Ong and Collier 2005).

An emerging approach has begun to grapple with the meaning of global change and transnational complexity in new ways by examining what have been labelled 'global assemblages' (ibid.). Global assemblages represent the tangible configurations through which global forms of techno-science, economic rationalism and expert systems gain significance and shape. The 'global assemblage' is the tool for the production of global knowledge, in the sense of knowledge about global forms, and knowledge that strives to replace socially, politically and spatially context-bound forms of knowledge. The term global refers to forms such as science, 'material technology', 'specialized social expertise' or systems of governance, whose validity rests on their impersonality and development without regard to context; their perceived ubiquity. In this context, 'global' forms are not everywhere, but present a capacity to continually reconstitute themselves through processes of decontextualization and recontextualization, abstraction, assimilation and movement, within diverse settings. The term 'assemblage' refers not to a locality in counter-opposition to broader, global forces, nor the site of hybridity in action, but to the articulation of evolving global forms into territorialized, emergent, evolving structures or domains that shape new ideas, technologies, collectives and discursive relationships. As a composite concept, the term global assemblage implies inherent tensions. Global implies breadth, homogeneity and mobility, and assemblage implies heterogeneity, contingency and context. Assemblages are not simply about resistance or assimilation, but about the relationships and activities that shape change in particular ways. I argue that the rise and, possibly, fall of biofuels provides an exemplar of an assemblage in action.

The frame of global assemblages provides an analytic domain through which we can understand evolving forms of organization and collective existence in the face of continual technological, political and ethical reflection, intervention and change. For example, we can look at the way in which biofuels as an emergent technology reconfigure the balance between food production, supply and consumption and interrogate the evolution of the complex or assemblage necessary for that to happen. Reflecting on the assemblage through which knowledge about biofuels is shaped and policies are developed and supported, and the implications of this for the systems in which biofuels interact, the development of technologies and ultimately societal outcomes provides a powerful analytical and critical tool.

This book will trace the global assemblages of expert knowledge, technologies, political and economic domains that shape and constitute the context in which biofuels are evolving, interacting and shaping complex interlocking systems. In doing so the book aims to delineate the contours of risk and responsibility that biofuels draw around the world. Biofuels risk profoundly shaping our relationship with nature and with agriculture, and altering the trajectories and priorities of global development, for good or ill. Understanding these dynamics and their implications is profoundly important from a material, developmental sense.

This book will contend that a confluence of conditions and drivers, the limits of our scientific knowledge, the demands of developed-country agriculture and the hopes of developing-country agriculture, increasing demands for secure energy provision, and a growing realization of environmental pressures, among others, have shaped an assemblage that not only drives and promotes biofuels as a simple solution to multiple problems, but generates significant risks as well. As the clamour for biofuels is global, so are the risks. As production of biofuels is by its very nature geographically constrained, and as the demands for biofuels mirror patterns of expenditure and income, so too do these risks. The assemblage of biofuels is global, and this represents not only the reach of biofuels and the circuits of knowledge, investment and demand that propel them, but also drives benefits and risks.

Philip McMichael (2009: 826) has argued that global investment in biofuels underlines the contradictions of the relationship between economic development and finite resources. He argues that a response to the increasingly apparent limits of what he terms the 'energy-industrial complex' that seeks to extend it into new economic and ecological niches is profoundly irrational, indeed 'perverse'. The 'corporate-driven substitution of fuel crops for food crops, the conversion of agriculture to a branch of the energy-industrial complex, deepens the fetishization of agriculture as a source of profit, rather than recognizing it as a source of life' (ibid.: 826).

It is, of course, only perverse if one accepts the contradictions between growth, development and the environment. If one does not accept these relationships as a contradiction but as an opportunity for further growth, and to generate new sorts of consumption, the logic of extending reaches is entirely rational, and this, too, shapes the assemblage.

This book argues that market and technological 'rationality' both masks and entrenches inequality and therefore vulnerability to change. This rationality is veneered by science, innovation and investment in the technology of biofuels as a solution to problems. Taken together, these shape, transfigure and migrate risks associated with investment in biofuels towards certain regions, countries and communities. These risks are either little understood, unacknowledged or deemed acceptable in relation to the perceived benefits. Part of the reason these risks may be deemed acceptable is because of who is at risk. A key concern is, of course, who makes these decisions. Ultimately biofuels, whatever their benefits, globalize risk. They build on existing inequalities, they extend existing contradictions, they draw on existing expertise and they perpetuate existing patterns of consumption. The global biofuel assemblage stabilizes these processes as rational and desirable and encourages significant changes in land use and livelihood for some (generally in the South) while perpetuating the conditions that generate inequality, expertise and political economy in the first place. Biofuels have been packaged as multi-purpose solutions to multiple problems, but one can argue that the assemblage uses biofuels to package unsustainability as sustainability, and uneven development as fortuitous.

Organization of the book

Chapter 2, 'Science', will focus on the history of biofuels. Where have they come from? How do they work? What do they promise? This historical perspective is important in seeking to identify biofuels not only as something new, to deal with tomorrow's problems, but as something historical which may provide insight, perspective and 'new' old knowledge.

Chapter 3, 'Systems', will explore our attempts to understand the potential implications of biofuel investment, and will underline the sheer complexity of doing so, the limits of our knowledge, and the ways in which policy-makers, in some respects, do not have the basis they might need to shape appropriate policy.

Chapter 4, 'Synergy', explores the political contexts that sustain and prompt investment in biofuels. A large part of the attraction is not born out of any obvious environmental implications, or concrete rural development outcomes, but out of a confluence of interests and politics. This is the allure of biofuels.

Chapter 5, 'Scale', will focus on some of the issues surrounding local context, appropriateness and prioritization. The global debate around the global need for biofuels risks obscuring some of the quite different ways in which they may operate in local contexts to provide local solutions. This chapter will explore the contestations inherent in context.

Chapter 6, 'Sustainability? The globalization of risk', the conclusion, will explore what all of this means. What does the rapid and contradictory evolution of a biofuel global assemblage tell us about our global priorities, our perception of the environment and our vision of the sort of world we choose to live in? Are we simply repeating recent history, fooling ourselves into thinking that we are preparing for the future? Are we accepting the lack of any limits to globalization? Finally, are we satisfied with shifting risk and responsibility around the globe and into the future without really thinking about the most profound implications of doing so? The book will conclude not by aiming to tie disparate strands neatly together, but by showing that disparate strands need to be teased farther apart, almost to breaking point, if we are to unravel the sorts of solutions we need for the 'perfect storms' we already face, and future storms we risk creating.

TWO

Science: biofuels, yesterday and tomorrow

Energy revolution redux

We stand on the brink of an energy revolution. Our reliance on fossil fuels will diminish as extraction becomes too costly and technically too difficult and we search for alternative energy sources that will not further exacerbate climate change. This is a profound moment in human history; modern civilization is in many respects the product of burning fossil fuels, using their combustion directly for heating and transportation, and indirectly to generate most of our electricity. This is the environment in which biofuels have gained traction over the past decade: an environment of shifting exigencies and new perspectives.

Biofuel refers to energy derived from biomass through processes such as combustion, gasification or fermentation (Demirbas 2007). These processes yield energy in the form of liquid or gas fuels. A range of biological sources can act as feedstock for these processes, including dedicated energy crops (such as grasses and trees), traditional crops (sugar cane and oilseed) as well as crop residues and degradable waste (for example, wheat straw, rice hulls and organic waste). The resulting fuel can be used in cooking, heating, electricity generation and transport (De La Torre Ugarte 2006).

Biofuels have in many ways been seen as a new technology, clean, efficient, natural, to replace an outmoded technology, dirty, unsustainable, dying. The term itself, 'biofuel', invokes something natural, 'from life', and some commentators have argued that 'agrofuel' might be a more accurate alternative (Bello 2009). The notion of something new, and better, replacing something old is powerful within policy circles. More specifically, the notion of moving from a finite, resource-based economy (dependent on increasingly limited fossil fuel) to an apparently limitless economy based on biomass has been identified as an important component of discourses surrounding the global knowledge economy (as identified by Castells) (Frow et al. 2009).

This notion of modernity is powerful in its imprecision, it appeals to many different interest groups, and seemingly offers multiple solutions to multiple problems. This contextless ubiquity is useful as a construct of policy but much less so when one begins to think about impact and practicality. This chapter will therefore introduce a series of country vignettes to explore biofuels, what they are, where they might go and, importantly, where they come from.

Biofuels – early history

> 'All of a sudden, you know, we may be in the energy business by being able to grow grass on the ranch! And have it harvested and converted into energy. That's what's close to happening.' (George W. Bush, February 2006)

Biofuels are touted as a fuel of the future. They will allow us to unshackle ourselves from our dependency on fossil fuels uninterrupted. We will be able to maintain our standards of living, our modes of transportation and our love of energy. Certainly, judging by burgeoning debate surrounding biofuels, in research reports, newspaper articles, projects and books, biofuels seem new. Countries have committed to setting targets for blending biofuels into petrol and diesel, entrepreneurs are searching for ways to unlock fortunes, and scientists and technologists are busy developing new, more efficient ways of tapping the possibilities of biomass.

Biofuels are, of course, in many senses nothing new. We have been unlocking the energy potential of biomass via fire for millennia, and indeed in some ways it is what defines us as human; consider the myth of Prometheus. For many in less developed countries burning wood or charcoal remains the primary means of providing warmth and cooking food. For many this source of energy is unlikely to change soon, and this speaks to our inability to affect global change and to deal with underdevelopment. Somewhat paradoxically, this underdevelopment in itself provides a rationale for new sources of energy, and a large component of the recent promise of biofuels has been the possibility of energy provision in poorer as well as wealthier countries.

The burning of biomass – wood – is clearly not new, but neither are biofuels. In 1912, Rudolph Diesel, inventor of the diesel engine, stated: 'The use of vegetable oils for engine fuels may seem insignificant today, but such oils may become in the course of time as important as the petroleum and coal tar products of the present time' (Nitske and Wilson 1965; Shay 1993). In 1925, Henry Ford, founder of the Ford Motor Company, and of assembly lines and mass production, told a *New York Times* reporter that: 'The fuel of the future is going to come from fruit like that sumac out by the road, or from apples, weeds, sawdust – almost anything. There is fuel in every bit of vegetable matter that can be fermented. There's enough alcohol in one year's yield of an acre of potatoes to drive the machinery necessary to cultivate the fields for a hundred years.'

Henry Ford was optimistic in his imagined energy returns and similarly optimistic hopes for biofuels are being repeated close to a century later. In the intervening period, however, first petroleum became the dominant energy source and then diesel was developed as the primary fuel for commercial diesel engines. Since then petroleum and its derivates have periodically been in short supply or prohibitively expensive, and this has led to the search for alternative energy sources, including biofuels. Kovarik, in his history of Henry Ford and the 'fuel of the future', notes that the reasons that led to the promotion of biofuels by Henry Ford were not simply about the most appropriate and effective fuel source. Ford, a committed agrarian, saw the promotion of biofuels as a means to engender a renaissance in American agriculture. An inventor who was a contemporary of Ford's, Charles Kettering of General Motors, worried about dependence on America's dwindling domestic oil reserves and therefore sought alternatives. The 'fuel of the future' of the 1920s was, as now, a technology shaped and driven by political context and a view of how the world should be (Kovarik 1998).

Biofuels 101

Biomass (including fuel wood and crop residues) is currently used by an estimated 2.4 billion people worldwide to generate energy

(Modi et al. 2006). As such, the notion of biologically based energy is not novel. Nevertheless, it has not previously been considered on such a wide scale and in such formal terms. While biologically based energy can take on various forms and be used for diverse purposes – such as biogas and solid biomass for heat and power generation – it is liquid biofuels for transport which appear to dominate interests (for instance, see WorldWatch Institute 2007). Biofuels are an attractive substitute for petroleum fuels as they have similar combustion properties and can work in existing technologies (such as combustion engines), and they can therefore be assimilated into existing supply chains with minimal technical adjustments. The production of biologically based transport fuel also offers potentially high profits from economies of scale. Indeed, biofuel must be manufactured in enormous volumes if current fuel replacement targets are to be met (WorldWatch Institute 2007; UN-Energy 2007).

One of the most desirable attributes of biofuels is that they can be so easily blended into petroleum fuels or relatively easily be used as alternatives to diesel fuels. Biofuels do not represent so much as a replacement for existing transportation fuels as much as an additive that enhances longevity. They can utilize existing refineries, power existing vehicles, lubricate existing transportation systems and extend current patterns of consumption and ways of life. In short, they can sustain the interests that currently own refineries, drive cars and consume resources. On one level, biofuels offer something new, the promise of a cleaner, efficient modern bioeconomy (Frow et al. 2009), while on another they offer stasis and perpetuation. For this reason biofuels have emerged as attractive solutions and opportunities to invest in across many different countries and contexts.

As we shall see later in the chapter, there are a number of locations where biofuels are already in commercial production, Brazil being the most prominent example, and we shall see later in the book that other parts of the world have been identified as places where biofuels can be commercially produced.

Biofuels are fuels that are directly derived from biological sources. Sources that lead to specific end products in biofuel production may be classified into different groups (or 'generations'), either 'first', 'second' or even 'third' or 'fourth' generation (for instance, UN-

Energy 2007). So-called 'first-generation' biofuels are in common use (OECD/IEA 2008). Many of their implications and interactions are documented in this book.

First-generation biofuels rely on food crops that boast readily accessible sugars, starches and oils as their feedstock. Ethyl alcohol, also known as ethanol, can be produced from any feedstock that contains relatively dense quantities of sugar or starch. The most common feedstocks are sugar cane, which accounts for around 65 per cent of all (bio)ethanol production, sugar beet, maize, wheat and other starchy cereals such as barley, sorghum and rye. These are currently the most popular sources of biofuel. Oilseed crops, such as sunflower, rape seed, soy, palm and jatropha (*Jatropha curcus*), can be converted into methyl esters (biodiesel) and blended with conventional diesel or burnt as pure biodiesel. Turning these crops into biofuels simply involves either fermenting the sugars or dicing up the fatty oils through transesterfication, the process used to convert oils and fats into fatty-acid methyl ester, which is commonly known as biodiesel because of its resemblance to diesel.

At the present time the vast majority of liquid biofuels in use are either biodiesel or bioethanol, although a much smaller amount of biomass-derived energy is converted into methane gas for use primarily in public transportation. While the amount of biofuel being produced globally is increasing rapidly it contributes only roughly 1 per cent of total road transport fuel consumption (International Energy Agency 2006), with about 85 per cent of this amount coming from bioethanol, and 15 per cent from biodiesel.

The potential of bioethanol and biodiesel to contribute to global demands for energy is limited by the availability of suitable land on which to grow crops and the relatively high cost and poor efficiency of most conventional conversion technologies. This potential is further constrained by competition for suitable land for food production. This constraint has led to two main approaches. One approach is to seek out alternative crops, the best known probably being jatropha, which do not need intensive management and can ideally grow in environments in which traditional food crops cannot (although, as we shall see, this is often not actually the case). A second approach has been to develop methods to convert cellulose (a component of the woody part of plants) into sugars that can then be converted to bioethanol. Two advantages of being able to convert cellulose is that plants with a high cellulose content tend to be able to survive on much more marginal land and that parts of plants not used for food consumption may be used.

Second-generation production technologies, in turn, rely on biochemical and thermochemical conversion. Biochemical conversion involves breaking down plants' cellulose and hemicellulose content into sugar molecules, which can then be fermented to yield bioethanol. Second-generation biofuels are derived from lignocellulosic materials as feedstock, among them dedicated biofuel crops such as perennial grasses such as switchgrass, trees such as poplar or willow and residues and wastes derived from agricultural production (Pin Koh and Ghazoul 2008). Dedicated energy crops are said to be suited to agriculturally marginal lands, thereby potentially enhancing biodiversity in the process (Tilman et al. 2006). Second-generation biofuels can theoretically be produced from a variety of non-food sources, which, again in theory, negates the problematic food versus fuel dilemma. The tough cellulose component of the feedstock is broken down into sugar and then fermented, either by enzymatic breakdown or acid hydrolysis, followed by fermentation. Alternatively, new thermochemical approaches, such as the Fischer-Tropsch process, where the biomass is gasified and then liquefied, are being developed.

The Fischer-Tropsch process currently represents the ultimate in second-generation biofuel production, primarily because it is far more energy efficient than current production techniques as it more completely reduces carbon in the form of liquid fuel. Hypothetically at least, the exothermic Fischer-Tropsch process is coupled with the endothermic gasification process and minimizes energy losses as carbon 'in' matches carbon 'out' (Moore 2008). The reason that we are not already using second-generation technology, and in so doing avoiding the pitfalls and trade-offs of first-generation, is the cost. Processing plants and refineries are expensive to build and maintain as they need to generate high pressures and temperatures. To be commercially viable they need to be large and therefore central, which means transporting biomass long distances for processing, which quickly begins to defeat the environmental purpose. Only a handful of commercially viable processing plants currently exist. Current research focuses on optimizing acid and enzymatic hydrolysis to address the former and searching for new strains of micro-organisms (such as bacteria, yeast and fungi) to facilitate the latter (Fulton et al. 2004; Hamelinck et al. 2005).

Third-generation biofuels focus on improving the feedstock. Developing oilier crops can greatly boost yields. Mining the genomes of crops may allow genetic agronomists to identify and alter the genes that control oil production. Developing trees with lower lignin content can make them easier and cheaper to process. An alternative approach is to identify entirely new types of feedstocks. Algal fuel, also known as 'oilgae', is a biofuel that may be generated from algae. Algae may be used as a feedstock from aquatic cultivation for the production of triglycerides to produce biodiesel; the processing technology is essentially the same as for biodiesel from second-generation feedstocks. A theoretical advantage of this approach would be to remove feedstock production from being land-based, which would cut down the potential for competition for land as biofuel production increases in the future. So far these technologies are theoretical, extremely expensive and very far off being commercially relevant. Even more theoretically, fourth-generation technologies hypothetically offer entirely custom-made feedstocks and microbes to process fuel. They will be discussed in more detail in the final chapter.

Brazil and bioethanol

Brazil has a long history of bioethanol production and is often touted as the most successful national producer of biofuels. Brazil's interest and investment in bioenergy production mirrored that of the rest of the world, with periodic energy crises, rises in fossil fuel prices and nationalist urges to secure access to energy stimulating innovation in research and policy. As early as the 1940s, Brazil was experimenting with using various oils and fats to combust engines. Neat vegetable oils such as castor seed, cotton seed and coconut were trialled (Pousa et al. 2007). The ban on the exportation of cotton-seed oil, which was the main vegetable oil produced in Brazil at that time, as a means to keep the price commercially viable and thus make it possible to use as a fuel for trains, perhaps represented Brazil's first foray into state intervention in biofuel production.

The main impetus for Brazil's biofuel programme was the global oil crisis of 1973, which caused the price of petroleum to rise in the 1970s and 1980s. In 1973, the cost to Brazil of oil imports was equal to one half of all exports. The increase in petroleum prices therefore exerted considerable pressure on the Brazilian economy (Goldemberg 2006). In response the Brazilian government embarked on two programmes: one, an effort to prospect for its own petroleum sources under its continental shelf, and two, a programme of producing large quantities of bioethanol from sugar cane as a direct substitute for petrol. This programme was called PROALCOHOL (alternatively known as PROALCOOL).

The programme was multi-pronged. Car manufacturers were encouraged to modify engines and vehicles to accommodate higher percentages of bioethanol blends. An infrastructure of bioethanol pumps in most of the country's petrol stations was developed. Initially, a variety of mechanisms were used to subsidize the programme: 'soft' loans to sugar-cane growers, encouragements to build bioethanol distilleries, and incentives to encourage people to purchase pure-ethanol-driven cars. The programme has been remarkably successful in scope. By 1980, no pure petroleum was being sold, only bioethanol–petrol blends. Initial bioethanol content targets of 5 per cent have progressively been raised over the three decades of PROALCOHOL's existence and now vary between 20 and 25 per cent (Pousa et al. 2007). Initial subsidies were reduced over time, and fuel prices liberalized, and bioethanol production is no longer subsidized. The development of efficiencies in the Brazil bioethanol system, albeit over a period of three decades, means that bioethanol–petrol blended fuel is cheaper than petroleum.

Total investment in the programme between 1975 and 1989 has been estimated at around US$5 billion. In contrast, it has been calculated that the avoidance of oil imports between 1975 and 2002 represented approximately US$52 billion. Over time, economies of scale and competition have led to a reduction in production costs from around US$110 per barrel in 1980 to US$30 per barrel in 2005

(Goldemberg et al. 2003). Agricultural yields have been driven up as agronomic techniques have improved and the area planted has increased. Significant amounts of investment in agricultural research have been a driving force behind this (Goldemberg 2006). A further cost reduction has been achieved by the use of sugar-cane bagasse (a by-product of sugar-cane processing) for energy production, which avoids the use of any fossil fuel in processing. This explains why the energy balance for sugar-cane bioethanol is so favourable, with up to ten output units for each input unit (ibid.). The importance of this issue is explored further in the next chapter.

Of all biofuels that are currently available, Brazilian sugar-cane bioethanol provides the most favourable greenhouse gas (GHG) balance. Brazilian bioethanol production can achieve GHG reductions of around 90 per cent compared to reductions of around 20 per cent for US-grown corn bioethanol. Investment in bioethanol plants that can burn bagasse and other waste products to provide energy for the production process significantly lowers GHG outputs. Continuous, long-term investment in research and development and refinements in agricultural practice and processing play a role, too. Highly efficient production processes coupled with high-yielding, highly appropriate growing conditions provide a context in which sugar-cane bioethanol production can thrive. Indeed, surplus bio-electricity, generated through bagasse processing, currently makes a contribution to Brazil's energy needs beyond that of bioethanol biofuel production. This is currently a relatively low 3 per cent of total energy needs, but plans are in place to significantly increase this contribution over the coming years.

Brazil's success is often presented as a model that other developing countries might follow (cf. Goldemberg et al. 2003; Goldemberg 2006; UNEP 2009), given that many of these countries already have significant sugar-cane plantations, which would allow domestic production and perhaps ultimately exportation.

There are, however, issues with Brazil's bioethanol sector. The potential of export, and Brazil's growing demand for energy, risks triggering rapid expansion of sugar-cane production and rapid land-use change. The amount of sugar cane under production, currently around eight million hectares, is projected to double by 2020, and output will more than double over that time frame. This creates risks for other types of land use in Brazil. The profitability of sugar cane risks arable land being turned over from producing other crops to sugar cane, which may in turn push the expansion of agriculture into the Amazon basin. This would trigger significant indirect emissions.

Other concerns revolve around poor conditions for workers and the threat of landlessness among the rural poor as investment in biofuels puts pressure on other land uses. Poor working conditions among people working on sugar plantations are not uncommon. Sugar cutters wield dangerous tools with limited safety equipment or training and are at significant risk from tropical diseases such as malaria. According to a report in the *Guardian*, 312 workers were reported to have died between 2002 and 2005, with 83,000 suffering injuries in the same period.[1] This compares with death and injury rates suffered in mining industries.

Brazil suggests that with careful nurturing it is possible to develop a sustainable sugar-cane bioethanol industry. Problems of expansion, exploitation and exportation may limit environmental sustainability and social justice in the future, however. This will be explored in more depth in Chapter 5.

Biofuels and the United States

The United States produces both biodiesel and bioethanol fuel, using corn (maize) as the main feedstock. Despite Brazil's efficiencies, described above, the USA has overtaken Brazil as the largest producer of bioethanol (Oxfam 2008). Together, the USA and Brazil account for around 70 per cent of all bioethanol production. Alongside bioethanol, biodiesel is produced in smaller quantities in states that sustain oilseeds.

Bioethanol has been seriously produced for fuel in the USA for around thirty years. The subsidy provided by the Energy Policy Act of 1978 launched the 'modern' version of the industry, and since then a succession of Acts have implemented a variety of instruments of support, ranging from partial excise-tax exemptions for blended fuel through to tax credits for the entity that blends bioethanol with petroleum (Tyner 2008).

Bioethanol production in the USA is heavily subsidized at other levels, too. Many states have complex combinations of state subsidies, renewable fuel standards and producer incentives. The total subsidy available for bioethanol in 2006 has been calculated to range between 28 and 36 cents per litre of bioethanol, or between 38 and 50 cents per litre of petroleum equivalent (Koplow 2007). No one would dispute that subsidies have made a substantial, long-term contribution to the bioethanol industry (Tyner and Quear 2006).

Import tariffs also support the US bioethanol industry. The USA applies a tariff of 2.5 per cent and half a dollar per gallon added duty (roughly 14 cents per litre). By comparison, the EU and Canadian equivalent import tariffs work out as €0.19 per litre and C$0.30 per litre (Oxfam 2008). Originally the tariff was enacted to offset the bioethanol subsidy, which applied equally to both domestic and imported bioethanol. While in some cases preferential access is given to domestic markets and therefore import subsidies are avoided, generally these tend to be countries that are insignificant producers of bioethanol. In the case of Brazil, most significantly, the USA, the EU and Canada do not give preferential access to their markets.

Given that Brazilian bioethanol, despite its own environmental and social implications, is more effective in terms of cost and GHG balance than its US equivalent, the protection of the US market is not rational from an environmental perspective. Typical GHG savings for US bioethanol are in the region of 20 per cent, and for Brazilian bioethanol are 90 per cent (excluding any effects due to land-use change) (WorldWatch Institute 2007).

Between 1983 and 2003 (apart from a couple of spikes) the price of crude oil remained relatively constant between US$10 and US$30 per barrel. This pricing, coupled with the fixed bioethanol subsidy, was conducive to growth in bioethanol production from 1,625 million litres in 1984 to 12.85 billion litres in 2004. Production grew by around 500 million litres per year during that period (Tyner 2007). Since 2004, the price of crude oil has begun to climb rapidly, often peaking at over US$100 per barrel. This high price, coupled with the fixed bioethanol subsidy (keyed to oil prices of US$20 per barrel), led to a huge growth in the construction of bioethanol plants. By 2007, bioethanol production was double the 2004 level (Tyner 2008).

The bioethanol boom was in many ways an unintended

consequence of the fixed subsidy being keyed to oil at US$20 per barrel. A further unintended consequence has been the dramatic increase in the price of corn – the main feedstock in the USA – up to US$230 per tonne in 2008. This rapid increase in the price of corn led to the planting of a further 6.25 million hectares, which in turn led to reduced soybean area and increased soybean prices (ibid.).

This rapid rise in commodity prices has stoked the food versus fuel debate. The price of corn rose from US$117 per tonne in early 2006 to US$233 per tonne at the end of 2007 – more than doubling in around two years. As corn is an important feed for livestock, particularly chicken, the cost of producing eggs and poultry meat increased by about 30 per cent by the end of 2007.

US production of bioethanol is a history of sustained subsidization and import protection, which owing to a disparity between bioethanol subsidy and the oil price has ignited in the past few years. This has occurred in parallel with the politicization of biofuel production in the USA. Former president George Bush, in his January 2007 State of the Union address, announced he was quintupling US biofuels targets.[2] These are not targets driven by environmental agendas; if they were then the importation of Brazilian bioethanol would be encouraged. Instead, they are driven by a desire to continue to grow economies, to sustain the grain belts of the US Midwest, and to secure a little more of the USA's energy supplies domestically.

Since 2007, bioethanol production has continued to grow. The 107 million tons of grain processed in bioethanol distilleries in 2009 was enough to feed 330 million people for one year. That was the year that 1 billion people were classified as hungry for the first time in history (FAO 2009a). In 2009, more than one quarter of the total US grain crop was turned into bioethanol.[3] Tyner, in an analysis of bioethanol subsidy in the USA, states that: 'We have entered an era in which agriculture supplies not only food, feed and fiber but also fuel' (Tyner 2008: 653). There will inevitably be struggles over exactly what agriculture supplies, and who will decide what is supplied to whom. In other parts of the world, beyond the Brazilian and US hold on the bulk of biofuel production, other countries are undertaking different paths, partly in an attempt to avoid food versus fuel issues, and partly to satisfy their own needs in their own contexts.

Jatropha and India

India has enjoyed one of the world's highest growth rates for several years. It has focused its biofuel policy on biodiesel, primarily derived from jatropha, which aims to meet 20 per cent of all demand for diesel beginning in 2011/12. Diesel is the main energy source for transportation, and India already imports a significant amount of crude oil as domestic production is limited. During 2004/05, the country imported 96 million tonnes of crude oil valued at US$26 billion, and an annual growth rate of over 6 per cent means that imports are due to rise to 166 million tonnes by 2019 and 622 million tonnes by 2047 (TERI 2004).

The risk of limited, or very expensive, access to energy putting the brakes on India's economic growth prompted the Indian government to launch a National Mission on Biodiesel in 2003. Jatropha was identified as the most appropriate crop given its high yields of oil, its ability to survive on less fertile land (and therefore compete less with food production), and ease of management and harvesting (Kumar Biswas et al. 2010).

The first stage of the Biodiesel Mission ran from 2003 to 2007 and sought to establish the viability of key activities within the value chain. Four hundred thousand hectares of land distributed across eight states were identified. India's highly decentralized bureaucracy means that many national and state institutions play a role in supporting the Biodiesel Mission. In fact, the Indian government has adopted a highly interventionist approach to the promotion of jatropha-based biodiesel production. The Indian government will initially take responsibility for almost all stages of biodiesel production, from the selection and development of high-yielding varieties of jatropha, the identification of appropriate land for planting, oil extraction, transesterfication and the production of biodiesel, to transportation and storage, and distribution to retailers. Pricing of seeds and biodiesel will be controlled and taxation and subsidization will be used to generate appropriate incentives for cultivators and other stakeholders within the value chain. India invested significant resources in promoting its programme.

The 2005 Mid-Term Appraisal (TERI 2005) identified several blockages in the growth of jatropha cultivation. First, local communities insisted on being given title to government-owned forest and wasteland before participating (Kumar Biswas et al. 2010). Second, private owners of marginal lands were not willing to grow jatropha unless they were given assured returns, backed by the government. Jatropha bushes have a gestation period of at least three to four years before they begin to produce seeds, and this means that a sizeable investment needs to be made by the cultivator before any return is experienced. Third, the unpredictability of market prices several years in the future also dissuaded cultivators from investing. The state therefore had to assure market prices and yields, and organize finance for investment across large tracts of land. A four-year repayment moratorium on loans, subsidized leasing of marginal tracts of lands, and guaranteed purchase of seeds have all been promoted.

Alongside issues related to perceptions of markets and risk, knowledge gaps raised other issues. The broader environmental risk of encouraging large monocultures is poorly understood, as is the risk of monocultures providing a ready environment for the spread of disease. Prospective yields for anything other than optimum conditions – real-world conditions – are simply not widely known, and this entails the possibility of exacerbating the risks described above as well as the state's guaranteeing of them (Altenbuerg et al. 2009). Initial results from Mozambique, for example, suggest that commercial jatropha production quickly drops off in less than ideal conditions, to the point where the levels of investment of time and resources necessary to manage jatropha orchards quickly become unsustainable.

Downstream, some private sector involvement, prompted by the Ministry of Petroleum, is discernible in the extraction and processing of seeds. Commercial biodiesel production has still not really begun as anything other than pilots. One of the key constraints here is pricing and appropriate subsidy. Production costs are slightly lower than in the USA or the EU but still higher than those for traditional diesel. Given the heavy intervention of the state throughout the production chain, a careful calculus of tax, rebates and subsidies needs to be worked out if incentives for cultivators are to remain and biodiesel is to be processed, produced and purchased. It may be difficult for the Indian government to extract itself from multilevel subsidization in order to leave a viable, sustainable industry. Given the limited experience of plantation growing of jatropha and its commercial production, the Indian government is essentially underwriting a large step into the unknown (Shukla 2008).

Despite these initial setbacks, demand for biodiesel in India, and demand for land for jatropha, is projected to continue to grow. If percentage-blending targets are to be maintained and energy provision is to keep step with economic growth, extremely large tracts of jatropha plantation will need to be planted. If by 2016/17 projected demand for diesel keeps pace with economic growth and the National Biodiesel Mission's target of a 20 per cent blend of biodiesel into conventional diesel is to be met, India will need to plant 14 million hectares of jatropha (Mandal 2005). This is an area roughly equivalent to the area utilized for all types of farming in the UK.

In theory India has ample marginal or 'wasteland' to accommodate the expansion of jatropha production to meet its own energy targets. In reality things may be quite different. Much land classified as 'marginal' is actually being used for some sort of economic activity, be it pastoralism or secondary livelihood activities such as gathering foodstuffs. Given, too, India's rapidly growing population and widening disparities of wealth, it is likely that many of these lands may already be informally occupied. Thus, there is a potential conflict between what may be considered a 'productive' use of marginal land and what may not.

Furthermore, uncertainties regarding how well jatropha will actually grow on marginal land, its maintenance requirements (apparently higher than initially thought), and the lure of state-subsidized and ensured profits may encourage production on less marginal lands, which may end up risking food production through competition for space, and of course the resulting changes in land use may have their own environmental and GHG implications.

Nevertheless, there may be other options. Unlike sugar cane/ bioethanol production, which exhibits large economies of scale (as we can see in the case of Brazil), biodiesel production may work as a small and medium-size enterprise (Openshaw 2000; Gonsalves 2006). Thus, there may be scope for a variegated, decentralized system of jatropha production that limits pressures on existing agricultural land and allows smaller-scale cultivators to benefit, too. A constraint to this more decentralized approach, however, may be that cultivators are reluctant to invest in jatropha production without market assurances and proximity to processing facilities, and developers of processing facilities may be reluctant to invest unless it is evident that feedstocks are readily available. The slow gestation of jatropha orchards, coupled with the greater organization and entrepreneurship required to provide enough local processing facilities, means that this approach, too, will likely be slow to evolve.

In some respects the Brazilian model is held up as an example for India to follow. Notwithstanding the different feedstocks, the technologies required and ultimately the fuel produced, the propellant of state intervention in a variety of ways and at a variety of points in the production chain is discernible in both cases. It is worth remembering that the Brazilian model took over a decade to become relatively sustainable, and bioethanol boasts the most positive GHG balance over petroleum-based fuels. The Brazilian model, too, is not without its problems. The state-sponsored large-scale land clearances used in Brazil to make land for sugar-cane production may be less palatable in India (Kumar Biswas et al. 2010). Regardless, the Indian government has invested heavily in its biodiesel programme and sees potential, even if as yet that potential proves elusive.

Tanzania, land and biofuels

In theory biofuels offer a real opportunity for African countries. Land and labour are cheap (Cotula et al. 2009), being located on the tropics ought to give a comparative advantage for biomass production, it may provide a rural development opportunity, and could help break many African countries' dependency on exported petroleum and help with balance of payments (Clancy 2008). Tanzania, for example, currently spends US$1.3–1.6 billion per year, some 25 per cent of its total foreign exchange earnings, on oil imports (Sulle and Nelson 2009). Economic growth is fuelling demand for energy consumption and raising the prices of existing energy sources.

Tanzania currently makes use only of approximately 25 per cent of its 44 million hectares of arable land.[4] Tanzania also has extensive tracts of land with low levels of rainfall and relatively poor soil fertility, which may provide the potential for jatropha cultivation. As of early 2009, somewhere between twenty and thirty-seven companies had requested land for commercial biofuel production, in areas ranging from 30,000 to 2 million hectares (ibid.; Kamanga 2008). This lack of precision with data reflects the constant flux of highly speculative investors as well as lack of government hold on what is happening in the country. Land speculation caused such a furore that in late 2009 the Tanzanian government imposed a moratorium on land sales until its national policy on biofuels is finalized. 'The Government was asleep' was the quote of the coordinator of biofuel production for the Tanzanian Ministry of Agriculture, Food Security and Cooperatives.[5] This routine of African governments struggling to keep up with the implications of new technologies is not unique, and has occurred before in East Africa over such issues as genetically modified crops (Smith 2010). States with limited capacity struggle to finely tune regulatory systems and policy instruments for the implications of new technologies, and biofuels are a case in point – and indeed it may well be that this perpetual state of 'governance lag' is not only a symptom of developing countries.

Tanzania currently produces jatropha and a small amount of palm oil (which has been produced since the early 1920s). Sugar cane is also currently widely produced, and while many proposals have been developed to diversify the use of sugar cane for biofuel production there have been few developments in this direction yet.

Unlike in Brazil and India, there is no state-led model for biofuel production. A mix of large-scale plantations, where companies control all aspects of production and processing, contract farmers and independent suppliers, whereby biofuel companies enter into contracts to purchase feedstocks locally, and hybrid models, which combine elements of the two other models, appear to be developing (Sulle and Nelson 2009). These models may end up evolving in different agro-ecological and climatic zones, and in relation to local access to land (Songela and Maclean 2008).

A number of recent biofuel projects capitalized through foreign investment have caused concern because of the large numbers of local people affected by alienation of their rights over customary lands (Gordon-Maclean et al. 2008). There is unease over whether land laws provide adequate protection against such alienation, and whether compensation payments provided for in the Village Land Act (1999) are sufficient to promote alternative livelihood opportunities (Sulle and Nelson 2009). There are real concerns over the risk of communities being displaced in favour of foreign capital in a country where the state controls all land.

Tanzania's apparent predisposition to selling its land is having another effect. Other countries, in seeking to secure their own access to food partially because of their own investment in biofuel production or to hedge against rising global food prices, are seeking to purchase or lease land. In 2009, South Korea tried to negotiate the acquisition of 100,000 hectares of farmland with the Tanzanian government, which at the same time was being courted by a United Arab Emirates company seeking a lease on farmland for rice cultivation to help secure food supplies for Gulf States.[6] Elsewhere in Tanzania foreign companies are growing sugar cane for bioethanol so that European countries can meet their EU biofuel blending targets (Mackenzie 2009).

There is a real danger of an international domino effect of food sovereignty (and insovereignty) developing, and this will be a particular danger for African states where land is cheap and ownership may be poorly regulated. Much as it is difficult to accurately apportion the impact of increased biofuel production on food price rises, it is equally difficult to categorize biofuel production as the sole or major driver of agricultural land acquisition. It is clearly a contributing factor, however, whether directly in terms of purchasing land for biofuel investment or indirectly in terms of hedging against rising food prices.

Many African countries, including Tanzania's neighbours, have already developed or are well into the process of developing sustainability principles or national plans for their emerging biofuel sectors. Tanzania currently has no formal national policies, strategies or regulations to guide biofuel development. A National Biofuels Task Force has been established, however, with the responsibility of promoting the development of policy on biofuels (Sulle and Nelson 2009). Subsequently, draft guidelines have been produced, but are waiting cabinet approval (as of early 2010).

The weakness of governance regarding biofuel development in Tanzania can be seen in limited planning, a lack of inter-sectoral coordination, and reactive policy provisions. This in turn has risked a lack of transparency in decision-making, a lack of regulation of foreign biofuel investment in the country, and a lack of consistent strategy regarding such investment and how to make the most of it for Tanzania. An Oxfam report states that:

> The emerging picture is one of investment for export with seemingly no requirements on companies to maximise value-addition within the country, supply national markets, form links with local companies, adopt production models likely to maximise opportunities for poor people, or work with local communities to increase access to energy. (Oxfam 2008)

Rather than a diversification of income and profit-making potential, Tanzania's experience seems to be an oft-repeated story of bending to power inequalities, global agricultural trade realities and missed opportunities. The case of Tanzania presents many of the compelling global implications of investment in biofuel production in miniature. Relatively powerless nations and communities bend to the will of international capital and in doing so open themselves up to new risks. Powerless governments struggle to develop policy frameworks to deal with new technologies, opportunities and the risks they pose. In this governance gap, capital and the private sector are unencumbered in their ability to reorganize land, livelihoods and production towards certain ends, and there are of course also risks in this. These circuits of capital and delineations of power seek to reorganize our relationships to nature, driven by new ideas and new technologies. In this we can see a migration of risk from North to South, and a concurrent migration of goods, whether they be tangible in profit or fuel, or intangible in the sense that we are doing something more positive for the environment, on behalf of other people. From this perspective, Tanzania may very well feel a sense of déjà vu.

The country cases presented so far give a taste of the range of issues, the scope of contexts and differences in technology that shape our current relationship to biofuels as both a policy idea and a technological solution. This already complex world of new technology risks even more – potentially unmanageable – complexity as current technologies may soon be obsolete as investment and policy imperatives drive the technological frontier forwards ever faster. There may be advantages in the adoption of new biofuels and new technologies, but as we see with the case of Tanzania, we have to be able to harness them, and understand them.

Later-generation biofuels

The preceding section has illustrated how important context, capacity and priorities are in shaping a country's experience with bio-fuels. The technologies underpinning the production of bioethanol or biodiesel are simple and almost ubiquitous. The levels and types of state planning and support are relatively more important than the technology (and possibly even the feedstock to a lesser extent) itself. It is often argued that so-called first-generation biofuels, of the types described above, are merely the warm-up for latter-generation, more efficient, more sophisticated technologies (Royal Society 2008). Latter-generation technologies may in some respects be much less dependent on context than the first-generation technologies we can examine today. The argument runs that they simply need sufficient investment and innovation in order to circumvent many of the perceived shortcomings of bioethanol, jatropha and the like.

Biotechnology may play a potential role in unlocking some of the challenges of 'next generation' biofuels (see, for instance, Fulton et al. 2004). It is a potential tool to accelerate advances in plant genomics and thereby the selection of high-yielding, less input-dependent energy crops. These may alleviate potential conflicts over land use and reduce greenhouse gas emissions associated with biofuel feedstock production. Biotechnology may also enable the engineering of energy crops that are pest-, disease- and drought-resistant, thereby guaranteeing a stable supply. Other traits that may be selected through biotechnological tools include rapid growth, low lignin content and the expression of enzymes within the crops themselves which can enhance the breakdown of cellulose (see, for instance, Sticklen 2006).

Biofuels provide a new opportunity for biotechnology, and private sector companies such as Syngenta and Monsanto have not been slow in showing interest. Using biotechnology to produce plants as feedstocks is likely to be seen as intrinsically less problematic than producing plants for food. Many of the biosafety issues, real or politicized, that continue to raise concern regarding the human consumption of biotechnologies are rendered moot. Indeed, biofuels may provide a possible avenue for biotechnology companies to gain traction in regions of the world in which they have struggled, Europe and Africa most prominently. It remains to be seen, of course, whether biotechnology can deliver, either in improving the yields of first-generation biofuel feedstocks or in developing second- or third-generation alternatives. The involvement of private sector agri-industrial companies does imply additional expenditure on the part of the farmer, and that may mean that small-scale developing-country farmers are locked out of new innovations, or need to make do with innovation originally developed with other applications, and sources of profit, in mind. It is likely that biotechnology will be no immediate panacea, regardless of the opportunities new markets may represent for Monsanto et al.

High oil prices, competing demands between foods and other biofuel sources, and the world food crisis of 2008 have prompted interest in the farming of algae as a source of vegetable oil, biodiesel, bioethanol and other biofuels. Particularly attractive characteristics of algal fuels include the obvious fact that they do not exacerbate competition for land, and their far greater potential to yield energy. Algal fuels may yield up to thirty times more energy per unit area than other first- and second-generation biofuel crops, although this is offset somewhat by their greater production cost per unit area.

Finally, other types of third-generation biofuels have been identified, for example bio-propanol or bio-butanol, which, owing to technical issues and limited production experience, are not thought to be practicable as fuels for several decades, although increased investment may change this (OECD/IEA 2008). Again, theoretically, the same feedstocks as for first-generational bioethanol can be used, but more sophisticated technologies can be utilized to process them. Propanol can be derived from chemical processing, such as dehydration followed by hydrogenation, for example. It seems that we have other generations of biofuel technologies to experience and experiment with before then, however.

The black box of biorefineries

Many of the benefits of 'second' and 'third' generation biofuels are encapsulated, quite literally, in the concept of biorefineries. According to the Royal Society (2008: 28) 'the objective of biorefineries is to optimize the use of resources, and minimize wastes, thereby maximizing benefits and profitability. The term biorefinery covers the concepts of integrating production of biofuels with higher value chemicals and commodities, as well as energy.' Biorefinery is taken to encompass a very diverse range of installations, different in approach and in size. A paper mill, for example, which burns waste lignin to provide heat and power, might represent a simple biorefinery.

Different types of processes, biological, chemical and thermal, may be integrated into a biorefinery, and ultimately a biorefinery may be able to sequester all of its carbon dioxide emissions, resulting in a fuel chain with a negative overall GHG metric. Much of this remains hypothetical at the moment (ibid.), although there is a certainty in progress: '[d]evelopments in biorefineries *will* lead to the use of lignocellulose as a feedstock with increasingly efficient conversion processes and use of wastes. Further development *will* lead to the breeding and growing of dedicated crops to optimize the production of energy and other materials' (ibid.: 28, emphasis added).

There is a strong sense when one reads the reports of organizations such as the Royal Society, or UNEP (2009), that science is certain, development is inevitable and we will unlock the insight needed to overcome current deficiencies in knowledge, process and technology. This type of rhetoric mirrors early debates surrounding the Green Revolution and the promise of genetically modified crops. When confronted with disappointment in both cases the technologies are reframed as stepping stones to something better that will solve the problems that were already touted as about to be solved (see Smith 2009). Intergenerational learning of past mistakes will almost certainly enable us to refine knowledge and solve problems. It is undoubtedly true that innovation is often built on making mistakes and learning from them; it is also undoubtedly true that learning from mistakes and making the next step is far from inevitable. History is littered with intellectual brick walls and technological cul de sacs.

There are two processes at play here. One is the presentation of a rational, linear teleology of innovation; that we can step by step inevitably solve any current shortcoming and look forward to a future of next-generation biofuel technologies that will cater for our needs. Implicit in this, of course, is a need to invest in current technologies for this to happen. This teleological perspective in effect means that no technological solution is ever a failure; instead there are, at minimum, crucial next steps in moving onwards. This reframing of knowledge and technologies into networks that stretch into the future can become problematic, as it makes it very difficult to assess a technology judged to be an 'early generation' technology on its own merits, and any benefits, and particularly costs, can get lost in accounting that includes benefits and possibilities that are yet to exist (Latour 1996). Investment in technology is locked in by the ability of policy to cast failure as success or setback as progress.

The lure of future, next-generation technologies also risks locking us into present technologies, which we are well aware do not meet our needs. The internal logic of surviving a painful transition from current, first-generation biofuel technologies in order to enjoy the fruits of next-generation innovation carries its own risks. First, we may never, despite our best efforts and most optimistic projections, develop next-generation technologies, particularly if investors and interest groups become too dependent on the status quo. This risk is already apparent in the USA, where legislation has positioned second-generation biofuels as a long-term supplement rather than a replacement for maize-derived bioethanol. Second, we need to consider whether, if investing in first-generation biofuel technologies can lead to second- and third-generation, it is the correct thing to do. As we shall see, many of the investments in land-use change, agricultural subsidy, research and development risk long-term implications environmentally, politically and economically, especially in the global South, where these implications will be most keenly felt, where there is less capacity to realign investment elsewhere as new technologies are developed, and where the comparative boon of the tropics is likely to be swiftly lost if second-generation biofuels do eventually take off (as enhanced biomass production is likely to be more than offset by the greater ability of developed countries to utilize new technologies that generate energy from agricultural waste).

We simply cannot see all ends, and cannot be certain that second- or third-generation biofuel technologies will not simply pose similar problems to first-generation technologies. It is not more efficient, more productive technologies which will cease to threaten food security and environmental degradation and promote monocultures, it is appropriate governance of biofuel technologies, whatever their generation, which is needed if that is to happen.

A second process that we can witness is the 'black boxing' of biofuels innovation. The notion of a biorefinery, which can achieve multiple tasks and possibly even sequester more carbon dioxide than it emits, is an attractive one. When one considers that many of the innovations needed for this to happen do not yet exist it seems less attractive. It is a powerful idea, however, as an object of policy-makers, lobbyists and scientists, and ultimately as an object for which any knowledge (or indeed existence) of the inner workings are rendered unimportant. The notion of a biorefinery is so intuitive and obvious that its inner workings can remain a mystery for the majority of us. The fact that we have little idea about how such refineries might one day work is not important because we are talking about the future and the promise of a tomorrow that we must invest in today. This further locks us into intergenerational investment in biofuel technologies.

From the past to the future

This chapter has sketched out both a history and a future of biofuels. The promotion of biofuels has historically been closely tied to the development of the motor car and the vagaries of oil prices. Early interest faded away as oil became plentiful and cheaply available, only to be reignited, initially in Brazil and the USA, as oil prices began to rise in the 1970s. More recently, a broader set of concerns around the environment, the decline of oil and energy security and, to a lesser extent, rural development have led to renewed interest in biofuels. Our present ability to efficiently produce biofuels is moot, as the next chapter will explore, but today's interest may pave the way for the promise of, and investment in, second- and even third-generation biofuels.

It is apparent, though, that context is key, whether that be historical, technical, political or environmental. Nation-states are presently investing heavily in technological pathways for which they have to be the innovators. It is difficult to see how Brazil's recipe for success can easily be replicated elsewhere. Biofuel production is intimately entwined in local agro-ecological contexts, and dependent on local capabilities to shape socio-technical systems to unlock the potential of bioenergy. Biofuel production appears tenuous and ephemeral, at risk of continual failure unless it is supported at every opportunity. Bruno Latour (1996) talks of the threat of new technologies disappearing before our eyes to be forgotten about moments later unless extensive networks of support exist to propel them forward. That appears to be so for all the cases described in this chapter, and these networks, too, cannot be simply re-created.

Path-breaking implies that little evidence or experience exists to chart a way forward, or map the implications of choosing a particular route, and this is apparent in the cases described here, to varying degrees. India is investing in a plant ostensibly chosen because of its capacity to grow in marginal lands, but is unsure how well it grows on those lands, and in some respects whether those lands should indeed be classed as marginal. The USA has invested very heavily in technologies that appear at best to be marginal in terms of their positive impact on GHG. Despite a long history of experimentation in biofuels, the USA is still working out its policy. And we have not even really begun to work out the implications of its policy for the rest of the planet. Tanzania has different issues to contend with. It does not have the capacity to replicate the highly interventionist, heavily subsidized strategies that have met with some success elsewhere, and struggles to develop the capability of the state to effectively govern international investments in bioenergy. It risks much, and in this lack of capacity it risks far more than the other countries discussed.

We find ourselves in a liminal state, barely cognizant of lessons from the past and excited by the promise of the future. This seeks to concurrently obscure the complexities and dilemmas of current investment in biofuels and justify it, in order to reach the future. We are promised that the second and third generations will render today's technologies obsolete, and that new sorts of platform technologies – paradigmatic technologies that will shape our future economic development – such as the integration of biotechnologies (Raguaskas et al. 2006) and even nanotechnologies for biofuel production will solve tomorrow's energy demands (see Royal Society 2008).

Of course, this cornucopianism is nothing new; we are regularly told in advance of future innovations, but the confluence of future, interrelated threats risks propelling investment, if not research, forward rather too quickly. In 2008, the UK's Chief Environmental Scientist, Robert Watson, called for a delay and time to think. He stated that: 'biofuel policy in the EU and the UK may have run ahead of the science'.[7] Indeed it may.

THREE

Systems: complexity and knowledge

Capturing complexity

Debate around and investment in biofuels have not just been prompted by the spectre of peak oil. Much of the attraction of biofuels lies in their perceived GHG neutrality. Not only can biofuels replace petroleum-based fuels, they can help mitigate climate change. As crops grow, they fix carbon from the atmosphere. When they are combusted as biofuels in engines this carbon is simply released back; the net impact on atmospheric carbon is, in theory at least, neutral. In reality, biofuels are not GHG neutral. There is a range of emissions associated with all stages of their life cycle, and a further set of emissions can be related to the implications of policy and practice choices associated with their use. Growing crops intensively, using nitrogen-based fertilizers, using petroleum-powered farm machinery, processing and transportation all require large inputs of energy, and that energy is usually fossil fuel. Of course, in reality biofuels simply have to emit less GHG than fossil fuels for them to be beneficial.

It makes sense, therefore, to assess the projected environmental impacts in thinking about policies that may promote biofuel production. If environmental savings through reduced GHG emissions are likely to be significant it may well be worthwhile investing large sums of money in subsidies or incentives to encourage biofuel production. Equally, if biofuel production is of only marginal benefit it may well be that the opportunity costs of investment do not add up.

Needless to say, arriving at an analysis that is both accurate and of use to policy-makers is exceedingly difficult. Technically, there are limitations and trade-offs between accuracy and accessibility in analysing the true environmental costs of biofuel production. Scientifically, we may simply not have the knowledge or the data we need to fully realize a workable analysis. Investment in biofuel production is currently outpacing good-quality research. Politically, interests, pressures and investments serve to shape the context in which analyses are undertaken and interpreted. Tim Searchinger, an important biofuel researcher, has stated it thus: 'A kind of reverse Murphy's Law in effect creeps into biofuel papers: if anything can go right, it will' (quoted in Oxfam 2008).

The optimism Searchinger infers has echoes of the earlier optimism of the Green Revolution, and of GM crops. Progress ought not to be tempered with undue pessimism, but it needs to be imbued with realism. This chapter will discuss some of the thorny issues surrounding analyses of the environmental impact of biofuels. That these issues relate as much to politics and the framing of issues as they do to technical shortcomings may be no surprise.

Setting boundaries

Recalling a sub-clause of the First Law of Thermodynamics, that energy cannot be created, is rather sobering when one considers the potential of biofuels. David Pimentel, a key critic of biofuels, has stated that 'the energy contained in all the vegetation across the whole of the USA only accounts for around one third of the total energy consumed in the USA per year' (cited in Moore 2008a: 9). This calculation, which sets the theoretical limits of the contribution of biofuels, is an exercise in boundary setting. According to Pimentel, the use of fossil fuels for transportation alone virtually matches the amount of energy sequestered in all US vegetation. Given that this sum takes no account of the massive energy inputs necessary to turn all of this plant matter into fuel, nor any account of the implications of doing so, biofuels appear likely to only ever make a vanishingly small contribution to meeting our future energy demands.

When one begins to refine these calculations through defining tighter (and more realistic) boundaries, such as assessing the energy sequestered in vegetation on farmland and the basic energy costs of producing the biofuel from this plant matter, for example, it quickly becomes apparent that meeting anything other than a fraction of current demand for transportation fuel is not going to happen with current technologies. This also puts the various targets (which range from a few per cent up to around 20 per cent, depending on country) for the blending of biofuels into petroleum and diesel in an overambitious light. Moore, in attempting to do some back-of-a-cigarette-packet calculations of his own in a recent article, stated that he 'learnt that many calculations do not so much attempt to fit reality as bend the mind of the reader to interpret them as reasonable' (Moore 2008a: 10). Oft-cited studies take no account, for example, of the energy needed to harvest biofuel crops and process the energy that they contain. They simply calculate the theoretical extractable energy within them. An example of this was quoted in an influential Royal Society report on the potential of biofuels (Royal Society 2008).

Use of biofuels appears intuitively sensible, since they unlock energy from the sun sequestered in plants to power cars, building on existing agricultural activity, and reducing greenhouse gas emissions into the environment. The reality is more complex, however, almost overwhelmingly so. We need more sophisticated analyses to temper our optimism.

Life-cycle analyses

Given the huge investment in and far-reaching implications of biofuel production it is not surprising that more sophisticated analytical tools have been developed to model impacts and assist with decision-making. Life-cycle analyses (LCAs) provide a means to measure the environmental impacts from 'field to wheel' of biofuels. LCAs are usually studied comparatively, in order to analyse which alternative energy source has the lesser environmental impact. More often than not these analyses focus on the impacts that production and use of biofuels might have on emissions of GHGs relative to emissions from the use of conventional petroleum-based transportation fuels (van der Voet et al. 2010).

A key concern in constructing life-cycle analyses is the realistic (and comparative) drawing of boundaries around the various inputs, outputs and processes that make up the theoretical 'life cycle' of biofuels production and use. Thus, a life-cycle analysis cannot be considered a neutral analysis of the environmental impacts of various types of biofuels; rather it represents what one would hope is a best guess at as complete a system as possible. As we shall see, neutrality can be forgone if one is intent on framing biofuels in a certain way, or making a particular point.

Analyses have been developed for so-called 'first generation' bio-fuels, as well as for latter-generation biofuels such as Fischer-Tropsch diesel, derived from lignocellulosic materials (e.g. agricultural co-products) and bioethanol from lignocellulosic biomass (e.g. switch-grass or woody crops). Nevertheless, the bulk of such analyses, and contention surrounding them, focuses on first-generation biofuels. The most striking feature of LCAs is the huge diversity of results. This can, of course, partially be explained by different biomass sources, different agro-ecological contexts, different processing techniques and other technical factors. Different methodologies, however, dependent on the different assumptions that underpin them, different (and new) datasets and advancing technologies, also serve to skew analysis.

One of the most striking facts in a systematic review of LCAs published in 2006 (Larson 2006), for example, was the fact that almost all LCA studies had been undertaken in a European or North American context. With the exception of a study on energy and GHG balances for sugar-cane ethanol in Brazil (Macedo et al. 2004), a similar study in India (Kadam 2002) and a study on biodiesel from coconut (Quirin et al. 2004), no other studies were identified. Palm biodiesel, which has been the focus of considerable interest in South-East Asia (Thailand, Malaysia and Indonesia) and latterly West Africa, had, by 2006, not been the subject of a systematic LCA. Nor did such studies exist for first-generation biofuel crops such as jatropha or cassava (Larson 2006).

This is quite bizarre given the global impetus behind biofuel production in developing countries. Studies in the European or North American context can provide indicative or proxy results, but given the context-specific variability and breadth of input values in LCAs – there is far more solar radiation in the tropics, for example – country- or region-specific studies are far more valuable in determining meaningful results.

Another striking conclusion from Larson's meta-analysis is the wide range of results reported for a given biofuel and originating biomass source. Within one major European study, the reductions calculated in terms of GHG per vehicle-kilometre (v-km) for rape methyl ester (biodiesel derived from oilseed rape) compared to conventional diesel fuel range from 16 per cent to 63 per cent. Soy methyl ester ranges from 45 per cent to 75 per cent. The range for bioethanol derived from wheat is shown to be anywhere from a 38 per cent GHG emissions benefit per v-km to a 10 per cent penalty relative to petroleum (CONCAWE et al. 2004).

The spread of results becomes even wider when conclusions from different authors are compared, which speaks to the impact of the range of assumptions and methodologies that may be employed. A study by Delucchi (2003) concludes that soy methyl ester will give a 213 per cent *increase* in GHG emissions per v-km when compared to conventional fuel powering heavy-duty vehicles. This is strikingly different from the 45 to 75 per cent reduction calculated in the European study above. It is pertinent to note that Delucchi considers a more extensive and detailed set of inputs for his LCA work than most other analysts, and this highlights both the difficulties of modelling highly complex systems and the extent to which assumptions, and the choices analysts make in including or excluding inputs, can skew results one way or the other. LCAs are in use today in regulatory processes to set environmental criteria and standards on biofuels. There have been calls for harmonized rules for LCAs, particularly in terms of assumptions regarding methodological issues (Menichetti and Otto 2008). We need to ensure we harmonize sufficiently sophisticated LCAs, however.

Dealing with uncertainties

LCAs for second-generation biofuels such as lignocellulosic energy crops are beset with other types of uncertainty. Field trials have been conducted for a variety of such crops, including willow and switchgrass, but limited yield data are available. These data may represent only a proportion of an emerging scenario if we assume that future yields ought to significantly increase as agricultural and processing practices continue to develop. Studies have, for example, projected switchgrass yields to increase over the coming decades in the same way that traditional crop yields have grown over the past hundred years or so owing to enhanced agricultural practices (Green et al. 2005). One would, of course, also have to factor in the development of improved processing technologies alongside simple improvements in yield. And one would also have to think carefully about the sorts of optimistic projections made for both future yields and future technologies given the recent history of overstating yield improvements in new cereal or GM crop varieties. Yield increases do not always translate well from controlled field trial to working farm. Finally, and possibly most importantly, one must consider the GHG implications of the increased inputs that will inevitably be needed to generate higher yields per hectare (Larson 2006). Generating these data for specific sites, where climate, soil, topography, agronomic practice, application of inputs, and distance from market, among other factors, will all play a part, requires considerable resources.

Two further factors related to discrepancies are the particular methodology followed for assessing N_2O emissions from fertilizers and the treatment of co-products in the technology conversion phase. N_2O emissions are particularly relevant for biofuels because of their relatively very high contribution to global warming; 1 kg of N_2O is equivalent to 298 kg of CO_2 emissions, so even small changes in N_2O emissions can significantly affect the overall GHG balance for biofuels (Crutzen et al. 2008).

Prospective, *ex ante* studies by definition deal in uncertainties and assumptions (see Smith 2007), and the wide range of results from LCA studies reflects this too. Methodological and analytical differences play an important part as well. LCAs are almost by definition narrowly focused on tracking cumulative emissions through fuel-type value chains in order to understand the implications of one-for-one substitution of one fuel type for another. LCAs typically do not analyse the consequences of policy actions or price changes. For example, if a biofuel could be produced more cheaply than a petroleum equivalent that price differential might lead to greater fuel consumption offsetting any environmental gains. To incorporate all possible feedbacks and interactions would of course be almost impossibly complex.

'Energy return on investment'

Energy cannot be manufactured. It is important to remember that all energy comes from nature, be it in the form of chemical energy stored in coal, oil or gas, or from energy sources such as sunlight, wind, waves or tides. Any process of obtaining energy and making it available for other uses is, in itself, an energy-consuming activity – for example, transporting biofuel feedstocks from field to refinery, or catalysing reactions to break down biomass. The concept of energy return on investment (EROI) is concerned with understanding these energy supply activities in context.

Any energy analysis must begin with primary energy capture at the boundary of the economy, but we may choose which subsequent activities need to be included. For example, we could consider the EROI of biomass supply. Or we could go farther and consider the net energy of bioethanol supply starting with biomass production and using some specified biomass-to-ethanol production process. Choosing how one analyses a system can have profound implications for the results of that analysis.

EROI is a simple ratio of (A) the energy supply activity and (B) the energy consumed by the activity in providing that gross output. It is important that B does *not* include the energy content of the primary energy resource that is being brought into the economy. That is derived from nature and is not part of the energy investment that we must make to exploit that energy source. For example, one would not include sunlight if one were assessing the EROI of solar energy as there is no 'cost' of supply of sunlight in relation to economic activity.

EROI is an important measure because it provides a gauge of the effort required to make energy available for purposes other than energy supply. It is a measure of the sustainability of an energy resource in making an independent net contribution to energy supply to the economy. Therefore, if EROI < 1, the source becomes a sink. But an EROI of only slightly above 1 can also be problematic, given other costs involved.

For example, consider an economy running on bioethanol with EROI = 1.34 (Shapouri et al. 2002). In order to supply one litre of bioethanol to the economy for purposes other than energy supply, the energy supply sector will need to produce 1.34/0.34 = 3.94 litres of bioethanol. Of those 3.94 litres, one litre will go to the economy, and 2.94 litres will go to the bioethanol production process itself (which, in this context, includes corn production). In other words, the energy sector itself would account for almost 75 per cent of energy consumption in the economy. This, of course, is not realistic. Bioethanol is not the only source of energy in the economy, and the energy (B) required to produce it may come from a number of different sources, other than bioethanol itself. It does illustrate that we need to understand the energy costs of a particular energy source as well as the energy derived from that source and its environmental impact.

The EROI is useful as it highlights the fact that, for some energy sources and fuels, very large energy inputs are required per unit of output, and that these energy inputs inevitably have to come from somewhere. That may not be a problem if the fuel in question has particularly desirable properties (for example, lower GHG emissions than alternative fuel sources in providing economic development, or energy security), but we must be prepared to subsidize it with energy derived from other sources. Indeed, this may be the case with liquid transport fuels. But this emphasizes the fact that the fuel in question (with its primary source) may not itself be a major contributor to energy supply, and that its availability depends on the existence of other resources with higher net energy. Even sugar-cane bioethanol in Brazil, despite its relative efficiencies, provides only a small percentage of the country's energy needs. From this perspective (and ignoring for a moment the impossibly large tracts of land that we would need to derive sufficient energy from biofuels for all our needs), the notion that biofuels can ever be anything other than a relatively small net contributor seems somewhat unrealistic. There are simply too many costs and trade-offs.

Flawed analyses

EROI is a useful measure as it allows us to compare energy investment, or input, versus output. Studies have focused, for example, on whether biomass-derived ethanol fuel delivers positive results – i.e. we get more energy back than we expend in extracting that energy – see, most notably, Farrell et al. 2006. This particular study, which essentially advocates further investment in biofuels as a viable energy source, makes assumptions that accumulate to skew their analysis in a particular direction. This led to a heated debate in the journal

Science, in which they published their research. In their paper, Farrell and his colleagues calculate an energy return for bioethanol of greater than 1 but smaller than 2. As Giampietro and Mayumi point out in their detailed study (2009), Farrell and his co-authors have chosen to ignore the initial input of primary energy in the process of bioethanol generation in their calculations. As discussed above, this is reasonable as there is no cost in the primary energy source. On the other hand, for their comparison with petroleum they have chosen to include the energy content of the source, crude oil, in their analysis of the effort used to produce it. As the discussion of EROI above states, the energy source as derived from nature need not be accounted for as an energy investment we must make to derive new energy. Essentially, Farrell and his colleagues are not applying the same set of assumptions to both analyses, and this renders their comparison almost meaningless (and emphasizes the efficiency of biofuels relative to crude oil).

Farrell et al.'s optimistic treatment of corn bioethanol produces an EROI of 1.2:1 (which does not account for externalities such as mechanization, labour or environmental impacts), implying that we would need to produce 6 units of bioethanol to net 1 additional unit for other purposes. Hagens et al. (2006) point out that devoting half of the USA's corn crop to bioethanol would require an energy input of 3.42 billion barrels of oil (almost half of the USA's current annual use) to net 684 million barrels of new bioethanol energy. This is before factoring in lost food production, soil nutrient leaching, ecosystem damage and water for irrigation. Again, this is not nearly efficient enough for large-scale utilization.

A further critique, by Kaufmann (2006), points out that the inconsistencies in the boundaries used in Farrell et al.'s study are so great that they imply that petroleum is the marginal fuel seeking to replace biofuels. If the data are as Farrell and his colleagues state, it is not rational that we have spent much of the last 100 years extracting fossil fuels rather than growing biofuels. Kaufmann concludes by asserting that once the boundaries of analysis are made equivalent, petroleum has a much higher energy surplus and a lower energy input/energy output ratio than biomass fuels. And: 'This result matches the economic reality described by the authors' first paragraph – biomass fuels, not motor gasoline, need subsidies and tax breaks' (ibid.: 1747).

All analyses and modelling of complex systems are based on manufacturing assumptions and proxies in order to aid comparison. Farrell et al.'s work undoubtedly exhibits analytical flaws – it is difficult not to introduce them when dealing with complexity and uncertainty, and highlighting particular flaws in an individual paper is not the purpose here. The purpose is to show that simple assumptions or errors can rapidly accumulate through calculations and entirely change the tenor of the analysis. Giampietro and Mayumi (2009) talk of the need to be very precise in defining the 'grammar' of energy analyses. It is also undoubtedly true that in talking about huge, multinational, heavily subsidized and politically charged issues it may be difficult to detach the science from the politics, or the equation from the ideology. Reflecting on EROI can be a useful counterpoint, however, as it obliges us to think carefully about boundaries between nature and the economy and how we define or draw them, and reminds us that biofuels in essence provide a new set of potential interactions between ourselves and the natural world that we do not yet fully understand.

Emissions and land-use change

The most significant omission from LCAs is the implications of land-use change. Land plays a major role as a carbon sink – sequestering a fifth of man-made emissions per year. As new land is put into production, significant GHG emissions can result, through, for example, cutting and burning trees or ploughing soil, releasing underground carbon.

Typical LCAs compare emissions from the separate steps of growing or mining feedstocks (whether corn or crude oil), refining them into fuel and burning the fuel in a vehicle. If only these process stages are analysed biofuels appear to match or exceed emissions for equivalent fossil fuels. But because growing biofuel feedstocks removes carbon dioxide from the atmosphere, biofuels in theory reduce GHG relative to fossil fuels. Analyses attribute credit for this sequestration effect, which is typically large enough that overall GHG emissions from biofuels are lower than those from fossil fuels.

Petroleum blended with bioethanol modestly reduces GHG emissions if made from corn and substantially reduces them if made from sugar cane (Searchinger et al. 2008). For most biofuels, growing the feedstock requires land, and therefore the credit represents the carbon-sequestering benefit of devoting land to that purpose. Searchinger et al. (ibid.) argue that most studies, in excluding land-use change, generally through assuming there is none, were one-sided as they included presumed carbon credits and ignored any potential carbon costs, i.e. the carbon storage and sequestration sacrificed by diverting land use. In order to produce biofuels, cultivators may plough up or burn forest or grassland, which releases into the atmosphere much of the carbon previously stored in plants through decomposition or fire. The loss of maturing forests or grassland also nullifies future sequestration gains as biomass grows each year, and this lost potential sequestration ought to be accounted for as a carbon debit. Farmers may instead choose to divert existing crops into biofuels, which indirectly causes similar emissions as farmers seek to expand cropland elsewhere to compensate for losses or to make maximum gain from increasing prices for increasingly scarce crops.

The majority of LCAs have operated on the assumption that any land on which biofuel feedstock was grown was previously 'set aside', which serves to negate any emissions from a change in land use. If in reality land was previously a carbon sink this ought to be factored into any analysis. For example, draining peatland and burning peat may lead to emissions of several hundred tonnes of carbon per hectare. Emissions from the destruction of peatland in South-East Asia have been estimated at about two billion tonnes of carbon per year (Royal Society 2008). The largest driver behind the destruction of peatland in Indonesia is the push to grow oil palm; the Indonesian government has recently stipulated that 40 per cent of palm-oil production should be set aside for biofuels (Oxfam 2008), driven by – mainly European – demand for biodiesel.

An analysis of the amount of emissions caused by clearing land compared with the emissions savings offered by biofuel crops is telling. Clearing Indonesian peatland and replanting it with oil palm would generate a carbon debt that would take 420 years to pay back (Fargione et al. 2008). Data like this do not generally find their way into LCAs and consequently have little or no bearing on policy-making. Decisions risk being made without fully mapping out the consequences, environmental and economic. Replacing peatland with biofuel crops, for example, is to all intents and purposes irreversible.

Cultivating biofuels on land that has significant carbon stocks is undesirable and counter-intuitive given the threats of climate change. Land that has the potential to sequester significant amounts of carbon needs to be protected. As rainforest is burnt so is the peat, and this means that up to thirty-three tonnes of carbon dioxide is released in exchange for every tonne of palm oil produced (Wetlands International 2006). In 2007 a UN report suggested that 98 per cent of the natural rainforest in Indonesia will be degraded or gone by 2022. Five years ago the UN predicted that this would not happen until 2032 (UNEP and UNESCO 2007). This accelerated destruction can in good part be attributed to demand for biofuels.

When one considers the case of the Tanzanian government recounted in the previous chapter and reflects on its inability to manage the influx of foreign investment surrounding biofuel development it is easy to see how thinking about the longer-term implications of development and analyses of assumption-ridden, Western-based LCAs are not particularly useful policy tools. Developing countries, many in tropical and subtropical regions of the world, tend to have rural-based economies (agriculture provides 80 per cent of Tanzania's population with a livelihood, for example; Sulle and Nelson 2009) that rely on natural resources. Foreign capital, too, tends to settle on natural resources, whether through tourism, agricultural exports, mineral extraction or now, of course, investment in biofuels. The governments of countries like Tanzania, Mozambique, Sri Lanka and Indonesia, to name but a few, need the knowledge, data and analytical tools to avoid making decisions that produce irreparable damage. Countries like Indonesia, Brazil and Malaysia contain vast resources of biomass and biodiversity. If they are to be destroyed or denuded in pursuit of biofuels we need to be absolutely clear why and what the long-term implications are. Many of these resources – rainforest, peatlands – are not going to come back once they are destroyed.

Opportunity costs: indirect land-use change

As briefly mentioned above, a further consideration in analysing the GHG implications of biofuels is that of indirect land-use change. The cultivation of biofuel feedstock has a significant opportunity cost. If a biofuel feedstock is planted on land that previously had another purpose an additional demand for land is created elsewhere. For example, replacing food crops with feed crops creates demand for the food crops to replace something else. The EU's demand for palm oil and the impact on Indonesian peatlands is a case in point. There are broader implications, too. Land-use change may displace people, increase food prices or permanently disturb carbon sinks.

Clearly, any assessment needs to be able to account not only for what type of biofuel crop is grown where, but also what type of land is being utilized, what the land was used for before, and what indirect land-use change may result from this elsewhere. This is tricky to quantify given the range of feedbacks and interactions involved. Delucchi (2003), for example, argues that linear LCAs need to be replaced with analyses that evaluate the impact of realistic actions – for example, the impact of a particular policy or intervention throughout a dynamic system. This, too, would be extremely complicated to undertake, but insights gained would be more useful than many of our current models. One might also argue that qualitative insights gained from dynamic models are far more useful and insightful than quantitative insights gained from linear, simplistic models.

The majority of previous LCAs have acknowledged but failed to account for emissions due to land-use change precisely because they are so difficult to quantify. Using an agricultural land-use model to gauge the impact on crops and cropland of increased use of corn bioethanol in the USA is insightful, and somewhat controversial.

In order to estimate land-use changes, Searchinger et al. (2008) used a global model to project increases in cropland in response to a possible increase in US corn bioethanol production of 56 billion litres above current projected 2016 levels. The model generated key factors that determine dynamic change to cropland, including fuel demands for corn, meaning that soybean and wheat lands switch to corn, impacting upon prices; as increased US croplands support bioethanol, agricultural exports decline sharply; and, when other countries replace US exports, they invariably have to plant more crops because of lower relative yields. The results of the analysis indicated that, when analysing indirect land-use change, corn-based bioethanol, rather than producing a 20 per cent saving in GHG, nearly doubles GHG emissions over thirty years, and increases GHG emissions for 167 years. Biofuels, derived from second-generation switchgrass, if grown on US corn lands, would increase emissions by 50 per cent. In comparison, second-generation cellulosic bioethanol may be able to make use of wastes that would not trigger land-use change (although the removal of significant agricultural wastes generates its own negative implications as they play an important role in sustainable soil fertility and structure as they break down). Bioethanol from Brazilian sugar cane, currently the most effective biofuel, could neutralize the up-front carbon emissions in four years if only tropical grazing land is converted. If displaced ranchers went on to convert rainforest to grazing land, however, the payback period could jump to forty-five years. Searchinger et al.'s (ibid.) analysis concludes by suggesting more complex analytical models ought to raise concerns about current US biofuel mandates.

Limited data and flawed accounting

If we are not careful, limited or problematic analyses creep directly into policy, as well as indirectly influencing it. The accounting mechanism used for assessing compliance with carbon limits in the Kyoto Protocol contains a crucial flaw that stems directly from the limitations or assumptions inherent within most LCAs. First, CO_2 emitted as a by-product of the burning of biofuels is not accounted for; and second, fluctuations in emissions due to land-use change which takes place as a result of the growing or harvesting of biofuels are likewise not taken into account (Searchinger et al. 2009). All bioenergy is treated as carbon neutral, regardless of the source of the biomass or feedstock. Any releases of carbon associated with biomass production – burning, for example – are simply unaccounted for.

One study, in particular, estimated that this error, applied globally, would create perverse incentives to clear land as carbon caps tighten.

A global CO_2 target of 450 parts per million under this accounting regime would encourage biofuel feedstocks to expand to displace virtually all the world's natural forests and grasslands by 2065. This would potentially release up to 37 gigatons of CO_2 per annum, which is of the same order as total human CO_2 emissions today (Wise et al. 2009). Another study predicted that incentives could encourage biofuel feedstocks to replace 59 per cent of the world's natural forest cover and release an additional 9 gigatons of CO_2 in the name of achieving a cut of 50 per cent in GHG by 2050 (Melillo et al. 2009). These land-use changes risk being incentivized because the assumption that bioenergy from feedstocks and biomass is carbon neutral encourages large-scale land conversion for biofuel production regardless of the actual net emissions.

The Kyoto Protocol caps the energy emissions of developed countries, but does not apply any limits to land-use or any other emissions from developing countries. Special accounting rules for forest management allow developed countries to offset their land-use emissions as well. Consequently, the CO_2 exemption for bioenergy use under the protocol wrongly treats the production of biofuels from all biomass sources as carbon neutral, even if the source involves obviously carbon-emitting land-use practices such as clearing forest for biomass in Europe, or converting it to biodiesel crops in Africa or Asia (Searchinger et al. 2009).

This error has been carried over into other regulatory frameworks, such as the EU's cap-and-trade law (2003) and the US Clean Energy and Security Act (2009). Both regulate emissions from energy but not from land use and exempt CO_2 emitted from biofuel production and use.

It is actually very difficult to suggest a simple approach to create better incentives. One could attempt to cap all land-use emissions and reward sequestration, although this would be exceedingly difficult to achieve, as it would be extremely tricky to measure all land-use emissions at national level, and then distinguish human-driven emissions from natural emissions. An alternative approach might be to trace the flows of carbon emissions and account for all emissions. Instead of working on the assumption that all biomass offsets energy emissions, biomass should receive credit to the extent to which its use results in additional carbon sequestration from enhanced plant growth (growing jatropha on previously unused land, for example), or from the use of residues or biowastes (using more efficient technologies).

This section has illustrated how assumptions around the science of biofuel production and its implications can translate into assumptions about how best to account for and incentivize particular types of activities. Errors in analysis and accounting may lead to errors in policy that generate perverse incentives towards inappropriate, or simply impossible, activities. This highlights the importance of the quality of the science, of data and analysis, and also the importance of encouraging debate and interaction between scientists and policy-makers, in order to ensure that policy is decided on strong premises. How, or whether, that policy is acted upon, of course, may be another matter entirely.

Complexities of certification

Alongside issues of how best to incentivize biofuel production and land use when so little is understood are issues of how, or indeed whether, we ought to certify biofuels to enable appropriate decision-making and encourage sustainable production – if indeed that is possible. Certification is potentially extremely difficult given the heterogeneity of possible impacts that biofuel production, processing and consumption may entail. Uncertainty resulting from the complexity of biofuel supply chains, uncertainties resulting from methodological and scientific issues (as detailed above) and uncertainties deriving from the differing and dynamic societal and environmental interactions and implications mean that effectively assessing the net impacts of biofuels production and use is fraught with difficulty (Woods and Diaz-Chavez 2007). This, of course, means that appropriate certification of biofuels is tricky. This is not simply a technical issue; governance will play a role too. Biofuel feedstocks are increasingly being grown in developing countries for developed markets, and therefore may cut across institutional weaknesses and international borders. Assessment and certification can be difficult in these contexts, ideas such as 'fair trade' being a case in point (cf. Ponte 2005; Giovannucci and Ponte 2005). In addition, biofuels markets are rapidly developing and expanding – both a problem for assessment and certification, and a rationale.

These emerging and evolving complexities provide the justification for a certification system. Such a system could be used to reward particular modalities of biofuel supply, based on their performance against a range of indicators. For example, it could provide a mechanism that could encourage improved productivities and efficiencies, and decrease impacts. There are uncertainties about the levels of detail and regulation needed to ensure improvements in biofuel supply chains, and the nature of the instruments and institutions needed to ensure that biofuels can meet supply demands without causing significant social and environmental change.

This chapter has already explored several of the key uncertainties in science, knowledge and methodology that drive the uncertainty regarding impact that besets the development of biofuels markets. These uncertainties need to be matched against principles that would underpin good practice and ultimately accreditation. These principles would themselves need to be broad and wide ranging given the complexity of biofuel value chains. For example, principles might encompass environmental aspects such as not destroying or damaging large carbon stocks (above or below ground), not destroying or damaging high biodiversity areas, not leading to soil erosion or degradation, contamination or depletion of water sources, or air pollution. Social aspects might include not adversely affecting workers' rights and working conditions, not affecting land rights or community relations, or not affecting rural livelihoods (Woods and Diaz-Chavez 2007). Clearly, these are very difficult principles to identify, assess and adhere to. Several are multifactoral – how would one disaggregate the destruction of a biodiverse wetland, for example, if there is more than one dynamic pressuring land use, or polluting? Likewise, understanding and attributing the interactions and pressures on livelihoods or community relations is complex and difficult.

Beyond the difficulties of understanding and attributing the impacts of biofuel production and development of markets are key questions about how to create institutions to govern biofuel development and limit impact. Expectations of workers' rights or environmental care vary widely in different countries and contexts. Environmental and social protection is weighed and policed differently in different regions of the world. These are complex and contradictory transnational issues. EU countries are the main motivators behind global governance of biofuel production, but other sets of EU policies and demands – blending targets, for example – are central drivers of biofuel production in developing countries. Internationally integrating drivers of demand and curbs on impact will be extraordinarily difficult given current complexity and uncertainty and future change.

Successfully integrating national and international markets and creating mechanisms to effectively govern them and mitigate risk will require new institutions (Mol 2010). These institutions will need new expertise and will have to deal with competing claims, interests and impacts. They will need to balance claims on resources that have thus far proved almost impossible to balance; we have a chequered history of managing impact and tempering demand for natural resources to grow economies.

Regardless of the efficacy of new institutions or the development of better methodologies to understand impact, some impacts, such as those on global, national and local food security, will be almost impossible to mitigate. They are driven by the interaction of markets, supply and demand that we have barely been able to control prior to the advent of global demand for biofuels. Access to food speaks to the heart of global inequity itself.

Connecting biofuels and food security

The relationship between biofuels and food security seems as impossibly complicated as the relationships between biofuel production, life-cycle emissions and land-use change. As the introductory chapter mentioned, relative attributions for the impact of biofuel production on global food price rises in 2008 range from 3 per cent (US Department of Agriculture), through 30 per cent (IMF and Oxfam), 60 per cent (OECD) to 75 per cent (Mitchell Report/World Bank). The huge range of these figures points to two things: first, that we are simply unable to systematically disaggregate and attribute cause with regard to changes in food prices; and second, that the factors driving changes in food prices are dynamic, variable and interact differently in different parts of the world. The matrix of causes of rice price rises in the Philippines, for example, is not going to be the same as for maize in southern Africa.

We have a somewhat problematic historical relationship to food production and the influence of technology. Writers like Karl Polanyi in his *The Great Transformation* identified the risks, especially to the poor, of agricultural mechanization and technological innovation, coupled with the 'vagaries' of the market. In 1944, he wrote: 'The actual source of famines in the last fifty years was the free marketing of grain combined with local failure of income' (Polanyi 1944). Thirty years earlier, Rosa Luxemburg (1913) analysed the tensions between capital and local production systems consequent on the incorporation of Asian and African peasantries into the late-nineteenth-century world market. These are not new processes triggered only by investment in biofuels.

More recently, we can reflect on the Green Revolution of the 1960s and 1970s which sought to deal with the threat of hunger in South and South-East Asia and elsewhere by conceptualizing a linear relationship between food production and food consumption, and a teleological perspective on how new technologies and approaches would sweep around the world through the sheer force of their own innovation, improving yields and livelihoods for even the poorest, most marginal farmers (Smith 2009).

The Green Revolution initially focused its interventions on issues of production, so-called 'isolable, technical problems which are so important that their solution would find acceptance and application' (Harrar et al. 1952: 25–6). This focus on seeds and not systems was problematic in its way, but also spoke to a belief that technological progress could usurp the power of agro-ecologies and of markets. Experience shows that while aggregate yields were raised, local consumption often was not.

Sen (1981), in his seminal entitlements approach, got to the nub of the problem by questioning why people were hungry instead of how they were hungry. The existence of food does not guarantee the consumption of calories any more than increased yields in field trials promise bountiful harvests in other parts of the world. Sen showed how people's abilities to access food were controlled by their ability to create bundles of 'entitlements' (used in a legal, not a moral, sense) which could be exchanged for food; entitlements could, for example, represent capital, the selling of one's own labour, or one's own ability to grow crops. Thus, the ability to access food should be understood as the result of the interaction of different systems, markets, social and political systems, and environmental conditions, for example. This points to the need for a far more nuanced understanding of food systems.

Alongside thinking about the ways in which individuals and households access food it is necessary to think about other drivers as internal or external to a country. A net food-importing country, particularly a low-income one, faces distinct challenges in terms of food price shocks leaving them particularly vulnerable. Clearly, the impact of food price rises on a net importer such as Haiti is quite different from that on the UK (Conceição and Mendoza 2009).

Internal factors tend to be exacerbated by poverty, and particularly inequality. GDP per capita growth in developed or emerging economies has outstripped food price rises over the past forty years or so. The opposite is true in low-income developing countries. Allied to this is the general trend of increasing inequality that we witness in almost every country, but especially in developing and emerging economies. The ubiquity of widening inequality means that we continue to see pockets of food insecurity in even the richest countries (FAO 2009a). While low-income countries are more predisposed to food price shocks, vulnerable populations in high-income countries, too, are at risk.

A related internal, and ultimately external, factor is that of changing patterns of food consumption. In emerging economies such as Brazil, China, India and South Africa (to name but a few), the growth of the middle classes means that more of different sorts of foods are being consumed. In China and India, in particular, a rapidly growing population coupled with a rapidly growing economy is creating demands for food, as well as other commodities.

External factors tend to be tied to international markets. International trade in food may be affected by volatility in commodities markets and financial speculation. The greater attraction of commodity futures to investors in the face of the decline in stock markets has meant that financial speculation is likely to have played a role in recent food price rises (Conceição and Mendoza 2009). Rising costs of the inputs necessary to produce food, such as oil or fertilizers (which themselves require oil to produce), also play a role in pushing the price of food higher.

Direct responses to the food crises exacerbate increases too. India and Vietnam, for example, restricted exports of rice, and the Philippines, a net rice importer, may have contributed to price rises by seeking to increase its own stockpiles at an inopportune moment. Other external factors to consider include environmental externalities linked to climate change and its effect on agricultural productivity. In Africa, for example, as far as we are able to forecast, climate change may result in a 17 per cent decrease in agricultural productivity, all other things being equal.

The impact of biofuels investment and production may be considered as interacting with both internal and external drivers of food price rises. Internally, significant investment in growing feedstocks at the expense or in place of food crops may increase a country's dependency on the importation of food. Externally, food crops such as maize, sugar and cassava can be directly used as feedstocks in biofuel production. This is the core of the food versus fuel debate; for example, in terms of global maize usage between 2004 and 2007, US biofuel production accounted for 50 million tons, while other uses accounted for 31 million tons. Since only 51 million tons of maize were produced during this period, these statistics imply there was a 30-million-ton decline in the global stockpile of maize (ibid.). This raises secondary concerns about our collective ability to respond to regional food crises; crises that are likely to be exacerbated by climate change and food price increases anyway. A further issue may be the land-use change discussed earlier in this chapter. Converting land from growing, say, food crops to growing feedstocks may have implications not only for negative GHG emissions; there may also be implications for food supply. We need to think very carefully about the structure of policies, subsidies and incentives that drive our global investment in biofuels in terms of these considerations.

In the short run, the roots of the 2008 food price crises may be related to unusually strong supply shocks and depleted stocks, which are to a respective greater and lesser extent due in some measure to investment in biofuel production. In the longer term drivers of food security remain entwined with the roots of poverty and inequality (ibid.). We need to understand better the interactions of short- and long-term drivers, and the role biofuel production plays in shaping them; and in potentially exacerbating poverty by impacts on food prices. Our current 'best guess' of a 3–75 per cent attribution of cause suggests we currently do not know enough.

Limits to knowledge, unlimited implications?

The leading edge of our innovation serves to expose the limits of our knowledge. Biofuels represent technologies that suture together systems, contexts and implications in complex and profound ways. Despite our most positive modelling, and invoking of multiple reverse Murphy's Laws, it seems very likely that biofuels – current generations – will be able to contribute little more than a fraction to fulfilling our energy demands. Our demands are too great and biomass is too limited, or at least needed for other purposes. This does not render biofuels unimportant, far from it, because the implications of our investment in them are out of all proportion to their potential contribution. Research shows that certain types of biofuel may take hundreds of years to pay back carbon debt before making a positive contribution to GHG (Searchinger et al. 2008). We risk losing for ever some of the most climatologically important peatlands in places like Indonesia where land clearing is taking place for palm-oil production. These landscapes will never be replaced. Even in regions of the world where particular types of biofuels, such as sugar-cane bioethanol in Brazil, may make a more obvious positive GHG return, we risk indirectly cutting swathes into the Amazon basin that can never be replaced, and we do not even know how to model these impacts.

It is important to reflect that the potential unsustainability of biofuel production discussed in this chapter is not unique. In many ways it simply reflects the general unsustainability of high-input, modern agriculture taken to the nth degree. Modern agriculture is not only fuelled by the sun; it is driven by the brute force of fossil fuels. Mechanization, fertilizers, pesticides and transportation are key constituents of modern, globalized agriculture, and this is not likely to change. If anything, the push for biofuels is likely to exacerbate fossil fuel use, and in doing so amplify the unsustainabilities built into agriculture. As the complexities and contestations relating to life-cycle analyses and incentives outlined above illustrate, we must be very careful about how we integrate biofuel production into our current patterns of land use. Biofuels must be adopted with a light touch, if they are to be of any benefit at all.

Ultimately, unsustainability is hard-wired into much of what we do and no amount of tinkering with analytical frames or boundaries will change that. Biofuel production will not solve the problem of the unsustainability of energy demands at no cost; it will simply send the bill elsewhere. The First Law of Thermodynamics, which states that energy can be transformed (from one state to another) but neither created nor destroyed, will loom large no matter what innovations, policies or incentives we choose to use.

This chapter has highlighted some of the complexities of agricultural and environmental feedback loops, and checks and balances. It has also highlighted the limits of our ability to adequately model them. These limits may be technical or conceptual, but we should not forget that they may also be political. Scientists are political actors too, even if hidden behind dizzyingly complicated analytical models or theoretical constructs.

We risk running ahead of ourselves. In our rush to replace oil we risk making investments and taking decisions that may be as irreversible as climate change increasingly appears to be. Our subsidies, securities and priorities shape choices the science cannot sustain or even provide insight into, and it is vitally important that we acknowledge this sooner rather than later. There is surprisingly little debate about the precautionary principle with regard to biofuels. The notion that we should not make a decision unless we are certain of any negative implications seems highly appropriate when we are thinking about decisions that will shape land use, livelihoods, the environment, capital and political will for generations. There is a sense of urgency out of all proportion to the immediacy of the issue, if not the scale, and it appears sensible to gather evidence, improve analyses and reflect rather than simply react.

The billions of dollars currently spent by the OECD on subsidies and support for the biofuel industry would be infinitely better spent on research and development into properly understanding the implications of first-generation biofuels and realizing the potential of second-generation biofuels. Yet the USA and the EU spend only a tiny fraction (about 8 per cent and 2 per cent respectively) of their biofuel investment on R&D (Koplow 2007; Kutas et al. 2007).

FOUR
Synergy: networks and interests

Assembling biofuels

The development of biofuels had a patchy start – initially seen as the primary fuel for personal transport à la Henry Ford, then overtaken by petroleum fuels, only to be more or less forgotten about apart from in distant parts of the world. As late as the 1990s, Cadenas and Cabezudo (1998) concluded that 'the future outlook of biofuels is beset with uncertainty'. Since the turn of the millennium interest in biofuels seemingly briefly ignited, only to be met with a new wave of resistance.

Biofuels as a technology, as an opportunity and as a contribution towards a better future, then, exist in an evolving, fraught and contested context. As the previous chapters show, there are many types of biofuel technology, multiple ways in which biofuel technologies are developed in different countries, and multiple interpretations of their efficacy, sustainability and impacts. Despite this diversity, the global idea of biofuels as an object of good policy, as something to be nurtured, has taken hold. In opposition, there are increasingly united voices of dissent, but it is not clear that these voices are decelerating progress or assisting policy-makers and practitioners to think through their actions and decisions more carefully.

We are faced with the unfolding of a global, socio-technical assemblage, which gives momentum, meaning and legitimacy to a biofuelled future. Trying to understand the nature of this transformation, from an emerging technology, through evolving policy to an evident impact, is useful as it allows us to unpick the processes, tipping points and ideas around which a new vision of the future will be articulated. The previous two chapters underlined our need to understand context, the reality in which biofuels might be produced, and understand complexity, both in terms of the multifaceted new sets of interactions biofuels may engender and of an acknowledgement of our limits – of knowledge and of our grasp and the reach of new technologies such as biofuels.

This chapter will explore the global perception of an emergent global technology and in doing so will highlight the ways in which technology, development and knowledge interact to propel us towards some future point. Implicit in this is a question of why, if biofuels are problematic, are we investing so much – capital, the environment, future pathways – in them?

Energy and (over)development

According to the 2006 *World Energy Outlook* (International Energy Agency 2006: 1), the global community is faced with two visions of the energy future: 'under-invested, vulnerable and dirty, and clean, clever and competitive'. In the same year, the then US president, George Bush, declared that we 'must break our addiction to oil'. Around the same time we witnessed a proliferation of targets – such as the United Kingdom's Renewable Transport Fuel Obligations (RTFOs), designed to stimulate the replacement of fossil fuels with more sustainable alternatives. Biofuels are one of the major renewable energy options considered in many, if not all, such targets. The EU is, for example, committed to replacing 5.75 and 10 per cent of its overall transport fuel supply with biofuel by 2015 and 2020, respectively (European Union 2007).[1] Similarly, the Environmental Protection Agency (EPA) of the USA set a renewable target of 7.76 per cent for 2008.[2] A number of developing countries have followed suit, Tanzania excepted, and countries such as Brazil, which had already invested heavily in biofuels, and India, which had set ambitious targets prior to that period, aim to invest more in the future. By 2007, at least sixty-four countries had developed national targets for renewable energy supply, including all twenty-seven EU countries (REN21 2008).

This is not surprising; our demand for energy is growing. Global consumption of marketed energy is projected to increase by 44 per cent from 2006 to 2035 (International Energy Agency 2009). The largest projected increase in energy demand is from non-OECD countries, such as India and China, which, with growing populations, thirst for development and economies growing year on year, risk stalling their growth if they cannot access sufficient energy. Historically, as was the case with Brazil and is the case with Tanzania, developing countries have been reliant on imports of fossil fuels. These imports often made up very significant proportions of all imports, and in relatively poor countries with weak and devaluing economies dependency on fossil fuels was a potential threat. Coupled with that, the price of fossil fuels has historically been very volatile. Three times in the past three decades oil-dependent economies have been affected by dramatic peaks in the price of crude oil – in the mid-1970s, the early 1980s and between 2004 and 2007 (UN-Energy 2007). Oil supplies are not seen as secure by the world's poorest countries. Abdoulaye Wade, president of Senegal, has stated that Africa's current oil crisis is 'an unfolding catastrophe that could set back efforts to reduce poverty and promote economic development for years'.[3]

The notion of a lack of access to energy as a brake to development is often cited, and provides a rationale for exploring energy options other than fossil fuels: 'The gradual move away from oil has begun. Over the next 15 to 20 years we may see biofuels providing a full 25 percent of the world's energy needs' (Alexander Miller, Director General-Sustainable Development, FAO).[4] Currently, biofuels account for only a negligible share of total energy consumption. In 2006, first-generation biofuels for transport, for example, contributed only 0.3 per cent of global final energy consumption, and represented 1.8 per cent of total transport fuels (OECD/FAO 2008). By comparison, traditional biomass accounted for almost 13 per cent of global final energy demand in 2006, the largest contribution of all renewable energies, which together accounted for 18 per cent.

Production of biofuels is increasing rapidly, however. Between 2000 and 2008, bioethanol production rose from around 18 to 67 billion litres of fuel per year. Production more than doubled between 2004 and 2008, and increased by 34 per cent during 2008 alone. This represents an accelerating rate of increase (REN21 2009). Over the same period, biodiesel production grew even more rapidly, albeit from a smaller beginning. Biodiesel production grew sixfold, from 2 billion litres in 2004 to just over 12 billion litres in 2008 (ibid.).

Other renewable sources of energy are growing alongside this increase in biofuels. Windpower, for example, doubled its energy-producing capacity between 2004 and 2008. Global solar photovoltaic capacity increased from 4,000 megawatts in 2004 to almost 17,000 megawatts in 2008. Over the same time period global investment in renewable energy grew from $20 billion in 2004 to $120 billion in 2008 (ibid.). Broad interest in renewables is clearly burgeoning. The theme of the 2010 World Bank *World Development Report* is development and climate change, and the report talks about 'energizing development without compromising the climate' (World Bank 2010: 189). The report lays out the problematic of the expected quadrupling of the global economy by 2050 against the impact of the associated projected doubling of GHG emissions over the same period. It argues that this trajectory is not inevitable:

> Solving the climate change problem requires immediate action in all countries and a fundamental transformation of energy systems – significant improvement in energy efficiency, a dramatic shift toward renewable energy and possibly nuclear power, and wide-spread use of advanced technologies to capture and store carbon emissions. Developed countries must lead the way and drastically cut their own emissions by as much as 80 percent by 2050 [...] But it is also in developing countries' interests to act now to avoid locking into high-carbon infrastructure. (Ibid.: 189)

Thus, the argument runs that we all must take responsibility for what is currently a legacy of development in developed countries. Developed countries have a responsibility to seek out new ways to generate renewable energy and capture and store emissions – which they do not have to do within their borders – and developing countries need to adopt new technologies in order to meet their energy demands and avoid falling behind.

Beyond oil

A focus on renewable energy for environmental and development reasons is laudable but there are other geopolitical considerations behind a push for biofuels. The dependence of a number of major crude-oil-importing countries – mainly the USA and those of the EU – on unstable oil-producing and exporting regions, notably

Russia, the Middle East and Venezuela, has triggered the former to develop programmes that seek to lower dependence on fossil fuel and increase energy security. This in some respects mirrors some of the reasons given for investment in bioethanol production in the USA in the 1920s by Henry Ford and his peers. Events in 2005 and 2006 underlined the over-dependence of the USA and the EU on exporters. Russia reduced oil exports to the EU on several occasions owing to disagreements with Ukraine on prices. Venezuela's president, Hugo Chávez, threatened to use the country's oil exports as a strategic resource. The continuation of the Iraq War and heightened tensions in the Middle East led to more problematic relationships between a number of OECD and oil-producing OPEC countries (Wirl 2009).

These geopolitical considerations, among other reasons, have led to increased volatility and far higher oil prices during the past decade. For much of the 1980s and 1990s, and up until 2003, oil prices were reasonably static, hovering somewhere between US$30 and US$40 per barrel. In 2004, oil prices spiked up to US$80 per barrel, then settled back down to US$60, before increasing rapidly to a high of around US$130 per barrel in mid-2008. Since then the price of oil has sharply dropped to levels closer to those of the 1980s and 1990s, owing mainly to diminished demand as a result of the global recession.[5] The historical high global price of crude oil has been a major stimulus to the production of biofuels, more so than other forms of renewable energy, as they can make use of much of the same production, processing and distribution infrastructure as fossil fuels. As we have seen, the oil crisis of the 1970s was the main driver behind Brazil developing its early sugar-cane bioethanol programme. Lower prices in the 1980s and 1990s pressurized the Brazilian programme, only for it to regain support as oil prices rose again (Mol 2007).

The sustained failure of agriculture

The sustained unsustainability of much OECD agriculture has provided a further stimulant to investment in biofuels. Overwhelmingly production-based subsidies leading to the overproduction of agricultural commodities, low prices, land being taken out of production (set aside) via other incentives, and lower (in many cases completely unsustainable) income levels for farmers have provided 'fertile ground' (to paraphrase an Oxfam Report) for the development of new markets for agricultural commodities. The USA and the EU, in particular, have deeply subsidized farmers and agribusiness to invest in the biofuel sector.

In 2007, support to biofuels among OECD countries amounted to somewhere between US$13 and 15 billion for fuels that provided a relatively meagre 3 per cent of transport fuel demand (Steenblik 2007). Of this support, approximately US$6 billion and US$5 billion are spent per annum by the USA and the EU, respectively. If one were to assume that current rates of subsidization were to continue, the 2020 blending target would cost European taxpayers over US$34 billion per annum (Oxfam 2008).

Support generally comes in the form of mandates that generate artificial demand for uneconomic biofuels; tariffs that protect domestic industries through limiting imports; and an array of subsidies and tax exemptions that provide support along the entire biofuel value chain, from production of feedstock, through to distribution and consumption. As the cases of Brazil, India and the USA, discussed in Chapter 2, illustrate, a vast array of instruments can and have been used to encourage innovation and investment in biofuels. Generally, these mechanisms of support have been targeted at supporting only domestic production of biofuels. This is changing, however, directly and indirectly. New forms of integration and relationship are developing that seek to ramp up production and use of biofuels, or offset some of the implications of their production. We can discern a shift away from domestic, state-organized biofuel production towards new integrated, international forms.

Actors, discourses and debates

Twenty years ago interest in biofuels was limited to isolated cases such as Brazil and a few examples in Africa, and interested parties were limited to people involved in those projects and possibly various sorts of engineers. More recently, certainly over the past five years, we have witnessed a global surge in interest and engagement. Most major players on the global energy markets, most environmental and development NGOs, and most countries and international organizations are busy formulating policies, strategies and perspectives on production and processing, supposed environmental benefits, and use of biofuels. Five years ago there was much more uniformity of opinion; biofuels were seen as likely to have a positive impact on GHG emissions as compared to fossil fuels, and this perspective was very powerful in drawing together what might normally be considered disparate interest groups – for example, the petrochemical and agribusiness sectors, governments and NGOs (Moore 2008a).

More recently the gathering of more empirical evidence and the development of more complex analyses have led to awareness of the complexities and problems of large-scale biofuel investments (Giampietro and Mayumi 2009). Now we can witness different actors developing a series of very diverse perspectives on the role of biofuels, some positive and focused on the potential of the science (see Royal Society 2008), some taking a more balanced perspective on the environmental implications (UNEP 2009), and some taking far more critical stances, claiming we should not be investing so much effort in a technology when we are not clear whether there are concrete benefits (see Oxfam 2007).

Yet, as some of the data presented earlier show, the evidence suggests that biofuel production has been accelerated just as interest groups are beginning to ask for the brakes to be applied. The 2008 US Food, Conservation and Energy Act is worth some US$290 billion and offers larger subsidies that those currently on offer for second-generation biofuels, provides loan guarantees to assist the development of commercial refineries for second-generation feedstocks, and incentives to encourage farmers to concentrate feedstock production around biomass facilities. The Act also slightly scales back subsidies for first-generation corn bioethanol.[6] This is an Act that affirms long-term investment in biofuel technologies.

In early 2010, the Obama administration reorganized US biofuels policy, including recalibrating targets to sustain support for first-generation corn bioethanols and biodiesels by incorporating them under the EPA's Renewable Fuel Standard to ensure that US biofuels targets for 2022 will be met. These calibrations include remodelling the environmental impacts of corn bioethanol (by updating data on crop productivity and yield improvement, and broadening its indirect land-use change model) to show that corn bioethanol is indeed a positive contributor in terms of GHG relative to fossil fuels.

The EU, or parts of it at least, have been slightly less gung-ho. In 2007/08 the Environmental Audit Committee of the UK Parliament released a report which claimed that the EU's current biofuel regime needed to be rethought because the environmental damage it caused far outweighed any benefits it brought in reducing GHG emissions. The report called for a moratorium on biofuels until a better understanding of the implications and interactions of supporting biofuel production was reached.

The EU Energy Commissioner, Andris Piebalgs, reacted strongly: 'On the contrary, the switch to [biofuels is] delivering significant greenhouse gas reductions compared with its alternative, oil.' He continued that the EU was promoting greater use of biofuels 'because this is the most immediately feasible way of significantly slowing the worrying growth of greenhouse gas emissions from transport'.[7]

Several other organizations and individuals have called for a moratorium on biofuels development, including NGOs such as Oxfam (ibid.) and ActionAid (2008), commentators such as George Monbiot, and, perhaps most publicly, Jean Ziegler, the UN Special Rapporteur on the Right to Food, who famously called biofuels 'a crime against humanity' in 2007.[8] Momentum, however, remains relatively unchecked.

Actors and networks

Manuel Castell's (1996) work *The Rise of the Network Society* captured something of globalization and the way it is experienced using the metaphor of 'flows' to describe new forms of movement that force us to reconsider our conceptualizations of time, space and power. Castells posited that flows and networks – rather than spaces and societies – are the architects of global modernity. The emergent institutional networks that shape society are driven both from within and between contemporary societies. This shaping requires expert knowledge that is able to exist between boundaries, of states, of disciplines, of other actors, and possessing this knowledge ties one into new populations of global elites who become key nodes within global networks.

Mol (2007) has begun to explore some of these ideas in relation to the emergence of what he terms a 'global integrated biofuel network', where 'environmental sustainabilities are more easily accommodated than vulnerabilities for marginal and peripheral groups and countries, irrespective of what policy-makers and biofuel advocates tell us' (ibid.: 297). In other words, these integrated networks act to incorporate or disincorporate knowledge that serves a purpose in support of the furtherance of the network. The idea that the role that networks play is shaping new knowledge, new technologies and new orders has also been developed by Latour (1996), who vividly describes the ways in which networks are central to the development of new technologies and their incorporation into society's fabric. Indeed, Latour sets out to show that the network, the in-constant-flux constellations of actors who interact with technologies, is in many ways far more important than any new technology in itself. For Latour and others (cf. Callon 1986), scientific activity and the pursuit of new knowledge (and the stabilization of new ideas in society) are ultimately about the ability of key actors, whether individuals or institutions, to build long chains or networks of association. Within these chains or networks, actors wrestle to secure strategic positions for themselves as obligatory points of passage. Occupying or controlling pivotal points in complex and lengthy chains and networks gives one the ability to 'shape' truth, or at least win arguments (Yearley 2005).

Instead of controversies, contestations or simply the back-and-forth sifting of complex situations being defined by who has the best access to the truth, the central idea here is that the truth results only from building a successful alliance. These alliances, it should be noted, are heterogeneous: they may be composed of actors, and institutions, of technologies and non-actors. A further analytical point is that the key claim of thinking of networks in this way is not that alliances are built up to override appeals to the truth, but that successful alliances constitute the truth in relation to whatever domain they are able to be influential in.

This offers a useful perspective from which to examine the evolving networks surrounding biofuels, or the emergence of a global integrated biofuel network (Mol 2007). Mosse (2005) has taken the core analytical approach and applied it to the formation of knowledge about international development, the creation of development policy, to good effect. He argues that development policy discourses become the end, rather than the means, of 'doing' development, because coherent and consistent development knowledge creates a far better framework for 'maintaining relationships' than contradictory and messy development realities. From this perspective, policy functions to mobilize political support and legitimize (rather than orientate) practice. Practice then becomes driven by the exigencies of organizations and the relationships they must maintain, and policy works well when it is possible to reinterpret practice as expressions of policy, and vice versa (ibid.).

Extending these explanatory ideas, or 'propositions' in the language of Mosse, further, we can begin to imagine how a 'global assemblage' around biofuels might develop. Thinking in terms of the interrelationships between expert knowledge and what we take to be 'expertise' regarding biofuels, the formation of new relationships between science and society driven by new biofuel technologies, and the economic exigencies and rationalities that subsidize, prioritize and propel biofuel policy and practice, provides a more critical perspective from which to think about the evolution of a global biofuel industry.

The politics of governments and governance

The resonance of biofuels with divergent interest groups is powerful indeed, and many of the interest groups are also exceedingly powerful. Up until a few years ago the main drivers of biofuel production were nation-states, countries like Brazil which hoped to offset the massive costs of importing crude oil, or the USA, which sought to diversify its energy sources and support the farmlands of the Midwest. While Brazil, the USA and the EU remain important actors, and powerful new nation-states such as India and China are becoming involved in biofuels themselves, the character of the biofuel assemblage is becoming increasingly cosmopolitan and transnational.

It is important to recognize that this transformation is not universal. As transnational organizations become more influential, new nation-states become involved in biofuels. Actors evolve and behave in different ways. Ultimately, the direct import and export of biofuels, or biomass intended for biofuels, is growing, alongside FDI. This renders the role of the once pivotal state less important. The shift from governments to governance is nothing new. The nature of the state is a constant topic of debate, particularly its role and continuing relevance in modern society (Jordan et al. 2005). The erosion of the ability of governments, which find themselves less and less able to deal with complex, increasingly global problems or respond to multiple, confusing and contradictory demands, has led to a governance-focused world (Stoker 1998).

There are two dimensions to governance. In the developed world, new divergent interest groups, new technologies and new globalized realities and problems are simply too vast and complex for solitary governments to deal with. In the developing world, notions of 'bad governments', an inability of resource-starved and capacity-poor governments to engender development, has led towards the notion of 'good governance'. The developing-country state is caught in the double bind of having to contend with the complexities of modernity as well. The state, then, if it is to function, must share its power through new forms of relationships and networks with non-state actors – NGOs, the private sector, multilateral blocs and other groups. Formal processes are replaced by informal processes, regulatory systems are superseded by participatory authorities, until ultimately non-state actors are allowed to organize and mobilize themselves. We can witness a transformational shift from the 'vertical' institutional order of domestically controlled markets and state strength to self-organizing networks and 'horizontal' coordination (Jessop 1998). Governance becomes an interlocking set of networks and social interactions rather than the political process of government institutions.

We can see in the global ordering of biofuels a broad shift from governments to governance. Perhaps we should not be surprised by this; the sorts of governance challenges that biofuels present, and the sorts of problems it is claimed they deal with, are precisely the kinds of issue that governance theorists claim governments cannot deal with. There is a risk in the disappearance of governments, no matter how gradual or imperceptible, and there is a particular risk in the disappearance of governments for those states, communities and interests which are not networked, do not have power, and do not have the capacity to quickly adapt. We must not forget that the biofuels global assemblage is shaped round expertise and interests, and they are qualities held tightly by those who also wield power.

The power of converging interests

Interests drive the biofuel assemblage and over time these interests have become more powerful, better organized and more integrated. Initially, farmers, cooperatives and processors were the main players in regional and national biofuels networks (Mol 2007). Today, large, multinational companies tend to be the big players. In the USA, agribusinesses like Archer Daniels Midland and Cargill donated US$365 million to politicians between 1990 and 2005, compared with US$182 million from oil and gas companies.[9] This is a good investment for Archer Daniels Midland, as biofuels currently account for 19 per cent of its profits.[10] Oil companies and car manufacturers are also powerful lobbyists, illustrating the potential of biofuels and their blending targets to lock in the profits of the broader fossil-fuel sector.

Agribusinesses like Archer Daniels Midland and Cargill are multinational businesses. Each is investing heavily in biofuel production outside of the USA. Indonesia and Malaysia, in particular, are favoured destinations for investment.

Corporations are also investing in biofuel research. Chevron is investing in research at the Georgia Institute of Technology, the University of California and Colorado University.[11] ExxonMobil gave US$100 million to Stanford University for biofuels research.[12] Shell, meanwhile, signed biofuel research agreements with six institutions around the world, including the Massachusetts Institute of Technology, USA, the University of Campinas, Brazil, the Chinese Academy of Sciences, and CoEBio3, of Manchester and Essex Universities, UK.[13]

These developments perhaps begin to explain the extraordinary growth of biofuels in the face of such ordinary evidence about their emissions and efficacy. An Oxfam report stated that 'biofuel targets

in rich countries are best understood as one part of a wide array
of support measures provided to domestic interest groups' (Oxfam
2008: 15). Martin Wolf, the chief economics commentator at the
Financial Times, introduced an article in 2007 thus:

> Energy security and climate change are two of the most significant
> challenges confronting humanity. What we see, in response, is
> the familiar capture of policymaking by well-organised special
> interests. A superb example is the flood of subsidies for biofuels.
> (Martin Wolf, *Financial Times*, 31 October 2007)

The multibillion-dollar per annum spend on subsidies for biofuel
production is testament to the effectiveness of well-organized special
interests. Philip McMichael (2009: 825) argues that 'responses to the
energy crisis follow a typical capital accumulation script – that is,
attempting to overcome barriers to profitability by extending the
realm of value creation, even as this intensifies capitalism's contradictions'. Thus a logical response to a crisis wrought by capitalism
is to encourage – via subsidy, policy or alliance – capitalism to seek
other realms of profit. The conversion of natural process into profit is
a profoundly political process, and historically generates new social
and ecological configurations that eventually uncover the inherent
unsustainability of the enterprise.[14] The biofuels assemblage, in the
guise of multi-purpose policies framed around sustainability, security
and development, seeks to legitimize the inherent contradictions in
the development of a global biofuel industry. By generating inter-
connections and leverage, the assemblage seeks to build networks of
fields of truth that allow the stabilization of investment in biofuels
as a rational policy response to an emerging set of problems. In
other words, the policy becomes the means to script the practice,
rather than to prescribe it.

Examining the ways in which new technological fields tend to
coalesce around problems is enlightening:

> All the companies which produce transgenic crops – Syngenta,
> Monsanto, Dupont, Dow, Bayer, BASF – have investments in crops
> specifically designed for the production of biofuels such as ethanol
> and biodiesel. They have collaboration agreements in a similar vein

with Cargill, Archer Daniel Midland, Bunge, transnational companies that dominate the global trade in grains. All this is creating new alliances. (Silvia Ribeiro, quoted in Padilla 2007: 6)

Here we see actors involved in deriving profit from other types of technologies seeking to engage with biofuels, as biofuels have ignited sufficient interest through their ability to frame multiple policy perspectives in a way that transgenic technologies have never been able to do. The narratives of transgenic crops and biofuels are in many ways similar and have their heritage in the Green Revolution, but the networks sustaining biofuels have unshackled themselves from the sorts of critique and perpetual references to the precautionary principle that so beset biotechnologies. Given that biofuels are in all probability far riskier in far more ways than transgenic crops were ever likely to be, this speaks to the power of the biofuels assemblage to generate its own significance and shape.

Sustaining agricultural research

Paradoxically, the growth of food insecurity, at least in part due to investment in biofuels, is creating new opportunities for biofuels research. Global agricultural science is massively funded; the global network of fifteen public, international agricultural research institutes known as the Consultative Group on International Agricultural Research (CGIAR) is funded to the tune of £425 million per annum (Alston et al. 2006). Five multinational companies – Bayer, Dow Agro, DuPont, Monsanto and Syngenta – between them spend approximately twenty times that amount each year on agricultural research (Leach and Scoones 2006). Several United Nations agencies play roles in governing and resourcing agricultural science, multilateral and bilateral donors contribute hundreds of millions of dollars per annum to research, and hundreds of national agricultural research systems interact with international agricultural research centres, universities and the private sector. On 11 November 2009, at the Global Food Summit, the director general of FAO, Jacques Diouf, called for an annual additional $44 billion investment in agriculture, including research (www.fao.org). The Royal Society of the UK has called for an immediate investment of £2 billion into agricultural science (Royal Society 2008), agricultural science is the fastest-growing research field in China (Adams et al. 2009), the USA has increased its 2010 agricultural science budget by 30 per cent after years of stasis (Stokstad 2009), and philanthrocapitalist organizations such as the Bill and Melinda Gates Foundation are also making massive new investments in agricultural science.

Increasing investment in agricultural research is in the main a direct response to growing concerns over food security. The increasingly global threat of food insecurity, outside of sub-Saharan Africa and South Asia, has piqued interest in investment in research. There is a real risk that developed countries may no longer be able to simply import themselves out of hunger in the future, and this is part of the thinking behind purchasing cheap agricultural land elsewhere to secure access to food. Agricultural research is going to be big business again.

Neither public nor private research is focused solely on food security. Biofuels are a huge focus. The Science Council of the CGIAR has called for care in further investment in first-generation biofuels, while at the same time calling for much more research in second- and third-generation technologies, and investment in the concept of multi-purpose biorefineries (CGIAR 2008). The CGIAR is careful to position itself as a key actor in this future research:

> As the CGIAR has a mandated, global responsibility for helping to reduce poverty and protect the environment through its research and research related activities, it is incumbent upon it to bring critically missing knowledge, practices or policies to bear on issues such as biofuels that directly or indirectly affect food security and agricultural sustainability in developing countries. (Ibid.: 5)

Several international research centres have been conducting research on first-generation biofuel feedstocks for many years. CIAT, the International Center for Tropical Agriculture, and ICRISAT, the International Crops Research Institute for the Semi-Arid Tropics, have worked on bioethanol processing from cassava. CIFOR, the Center for International Forestry Research, has been focusing on the potential of forest-based bioenergy for climate change mitigation. CIMMYT, the International Maize and Wheat Improvement Center, has been assessing and prioritizing investments in speciality maize as a biofuel feedstock. ICRISAT has been working on developing alternative feedstocks, such as sorghum and pongamia. There are potentially tensions between the CGIAR's mandate of reducing poverty and protecting the environment and engaging in biofuels research. Nevertheless, international agricultural research is gearing up to conduct experimentation on new-generation biofuels and the implications of current-generation biofuels.

Philanthrocapitalism is playing an increasingly influential role in agricultural research, especially in developing countries (Bishop and Green 2008). The Bill and Melinda Gates Foundation is investing in the development of algal biofuels, and has provided funding to Stanford University to assess the socio-economic implications of investment in biofuels. The Foundation's ongoing support of research into higher-yielding varieties of crops is a less direct way in which biofuels research is supported, either in terms of potentially enhancing feedstock yields or partially avoiding food versus fuel demands for biomass.

The international infrastructure of agricultural research is responding to rising food insecurity, new technological frontiers and new funding and investment possibilities. The private sector, the public sector and, increasingly, relationships between the private and public sectors are setting new research agendas that directly or indirectly engage with the potential of biofuels. As first-generation biofuels appear less and less attractive, research and investment are beginning to look to the future and to new generations of technologies.

We are witnessing something of a biological dot.com-style boom. Investment in biofuel start-up companies is rapidly growing. These companies tend not to focus on large-scale crop-based biofuels (which are of most relevance to the developing world), but instead on high-technology, high-value approaches. Venture capitalists are increasingly focusing on 'platform technologies' such as algal approaches or synthetic biology, which require less capital than large-scale approaches, and potentially produce higher-value fuels such as butanol (Luxresearch 2010). Approaches such as attempting to produce cellulosic bioethanol, or using single microbes to break down cellulose or ferment sugar, thus cutting costs, are commonly focused on. Concentrating on algal technologies is higher risk given relatively untried and untested technologies and uncertainties about whether commercial viability is even possible; nevertheless there is much interest in this field too. Many of the companies involved, such as Canada's Logen, are supported by petrochemical companies such as Shell; others, such as Qteros, are backed by private financiers such as George Soros. The most common exit strategy for these companies is to seek partnerships or buyouts with petro- and agro-chemical companies such as Shell, Chevron, Monsanto or Syngenta (ibid.). In this we can discern other sorts of relationships and synergies developing between emerging technologies and established concerns.

While the science suggests that future generations of biofuel technologies hold the most promise and present the fewest problems, we are still investing heavily in first-generation technologies, and not just for sound scientific or technological reasons. While scientists and investors look to the future, policy-makers look to the present.

Sustaining unsustainability

OECD agriculture has long been in crisis. The size of current subsidies and, one could argue, the headlong rush into biofuels by Europe are testament to that. The distortions caused by mainly US and EU agricultural subsidies have had long-term effects on the profitability and sustainability of agricultural production elsewhere (see Oxfam 2001). The size of subsidy necessary to support OECD agriculture has become unsustainably large, and investment in biofuels provides an opportunity to reframe what subsidies are for. Casting subsidies as promoters of global environmental sustainability, energy security and sustainable agriculture is far more palatable, and appeals to far more people, incorporating far bigger networks, than simply 'sustaining US or EU agriculture' ever did. Unsustainability is hardwired into agricultural subsidies by definition, especially as they almost inevitably serve to render previously sustainable agriculture elsewhere unsustainable as well. A tripartite rationale essentially divides the cost of unsustainability by three, making it somehow seem less expensive. In much the same way that capitalism looks to exploit new opportunities in the face of crisis, one might argue that so does the subsidy. It is quite difficult to tell whether the relationships between capital and nature, and subsidy and nature, are different from each other in terms of their internal logics or ultimate effect.

Mol (2007, 2010) argues that while biofuel development was initially a state-led, national project at the turn of the millennium, a global biofuel system started to emerge a few years ago. At that point 90 per cent of biofuel production was consumed domestically, but that is rapidly changing (Dufey 2007). Imports of bioethanol into the USA tripled between 2004 and 2006,[15] exports of Brazilian bioethanol increased enormously, and trade in biofuels increased in many other countries (Junginger et al. 2008). The most well-developed national biofuel systems, those of the USA and Brazil, have until recently been protectionist, with high tariffs for imports, but the signing in 2007 of an 'Ethanol Accord' between President Lula of Brazil and former president Bush of the USA signalled the beginning of a notional globalization of biofuels. The Accord links together the producers of 70 per cent of the world's bioethanol, and promises more coherent bioethanol policies between the two countries.

Many OECD countries will simply not have the domestic capacity to meet national demand, particularly in light of ambitious targets such as the EU's goal of 5.75 per cent biofuels blended into transport fuel by 2010 and the USA's goal of a 10 per cent blend by 2020.[16]

Indonesia and Malaysia are already expanding their oil palm plantations to meet this growing demand. Together they hope to supply up to 20 per cent of the EU market (Dufey 2007). Brazil and other Latin American countries (notably producers of palm oil such as Ecuador and Colombia) are looking for export opportunities, and countries in Africa and Asia are investing rapidly in large-scale production of plants such as jatropha.

There are knock-on effects of the globalization of biofuel production. Developing countries risk becoming locked into producing for target-driven EU markets. Jatropha, for example, takes several years to become established and, without the promise of guaranteed seed purchase, as is the case in India, there is little incentive to invest in it for domestic markets. Given that the EU is currently reviewing its future blending targets in light of concern over the real environmental savings of biofuels there is a risk that markets may just disappear (Moore 2008a; Smith 2010).

Investment in biofuel production means potentially competing with land for food production, and there is an extensive history of the implications of African countries getting caught up in export agriculture at the expense of domestic food production. New hierarchies are being created. Less developed countries which have invested in biofuel production are turning to cheaper agricultural land in Africa as a means of ensuring their food security. There are obvious implications for food production and sustainability here. This will be covered in more detail below.

Finally, creating an international market for biofuels means thinking about issues such as regulation and environmental auditing (Mol 2010). This may, in some senses, provide an opportunity to govern biofuel production more effectively, but it may also introduce new distortions, instabilities and unsustainabilities into the assemblage. Policy instruments such as certification of labelling may work relatively well in cross-boundary systems, and such instruments may oblige national biofuel systems to account for the environmental, food security and labour implications of biofuel production. Brazil has introduced the 'Social Fuel' label, which aims to promote social inclusion along the value train (Dufey 2007). As discussed earlier, there are broader issues of governance and sovereignty at play here, however.

At an entirely different level, the globalization of biofuel generates demand for multilateral liberalization and an end to protection (Mol 2010). Other private forms of governance have been developing. Multinational companies are developing corporate responsibility policies and coordinating international value chains; international networks encompassing different sets of actors, such as NGOs, are forming to lobby and pressurize in certain directions (Verdonk et al. 2007). These emerging networks of actors may provide useful oversight with regard to the environmental aspects of biofuels, although intervening indirectly in issues such as the purchase and appropriation of land for the production of feedstock or food may be difficult.

There are expectations that the World Trade Organization (WTO) will engage with the evolving global biofuel market through its remit for agriculture (Howse et al. 2006; Motaal 2008). This may be problematic, as most biofuel-producing countries are convinced that import barriers need to be taken down, but are equally convinced of the need to subsidize their domestic producers, processors and users. Indeed, it is still not certain that any domestic agricultural sector can sustain itself without significant levels of state support, at least for an extended period during the growth of the sector. The WTO might be a problematic governing institution in other ways; mandatory, and universal, regulations and standards would be very difficult to pitch given the diverse set of countries and actors engaged in the biofuel sector.

It is difficult to identify in the current institutional matrix just who should take responsibility for governing the market for biofuels. Global food security is best served by stable agricultural commodity prices, and these prices need to be pitched to allow access to food for the world's poorest people, and encourage investment in increasing global agricultural production.

Mol (2010) suggests that the increasing interrelatedness of agro-food and energy systems means that stabilizing agricultural food prices cannot be done without also stabilizing energy prices, and current global institutions are ill equipped to do so, and thus risk ensuring food security. The WTO's focus on liberalization is narrow, and liberalization and deregulation will not provide conditions for price stability. In a similar vein, the current sectoral focus of multilateral organizations, such as the UNEP, UNDP, the World Bank and UNCTAD, risks breeding complexity rather than coherence. As we enter the second decade of the century, and really the second decade of concerted, multi-country investment in biofuel production, we witness the ungovernability of biofuels as the biofuel global assemblage extends and increasingly ignores national boundaries. Averting future food-price spikes akin to those of 2008, and ensuring the environmental sustainability of complex, international interrelationships, may be a long way off. Finally, as we shall see, sensible governance of the international land-use changes wrought by biofuel production cannot wait.

Reconfiguring global land use

The power of capital, interests and subsidy fuel demand for land, and land is at its cheapest and people's rights to lands are weakest in Africa.[17] The logics of capital push for the exploration and exploitation of new horizons, and Africa, not for the first time, is the focal point for biofuel-driven land-use change (Cotula et al. 2009).

Since around 2007, the acquisition of large tracts of farmland in Latin America, Central Asia, South-East Asia and particularly Africa has accelerated. Land which had previously been of little interest is now perceived as a development opportunity. As food prices rise and the amount of land under crop production tightens, private investors have bought up land, governments have sought to acquire land elsewhere in order to secure their food supply, and recipient countries have, more or less, welcomed the opportunity to attract investment.

There are several broad implications of the globalization of agricultural land ownership and the sites of production. First, the balance between large-scale and small-scale agriculture is being shifted, with particular implications for small-scale farmers and those who are sustained through their production surplus. Second, the relative importance of export agriculture is likely to increase, which again removes food from local markets. Third, the role of agribusiness and the vertical integration of agricultural production, processing and distribution becomes ever stronger (ibid.).

Typically, there is as yet little research regarding the broader implications of these global shifts in land ownership. As we can see in relation to trying to understand the role biofuel production played in recent food-price rises, these are complex, multilayered systems. We need to be able to understand the impacts on, in particular, global staple food prices, local livelihoods and developing-country food security. We also need to understand the continual strengthening of agribusiness and the implications of this on state policy and food security (at all levels) and the technological and environmental implications.

Large-scale land acquisitions need to be understood in the context of growing economic relationships between the developing and developed world. Economic liberalization, the globalization of transport and communications, and global demands for food, all year round, coupled with new global players such as China, India and Brazil, have fostered foreign investment in many parts of the global South – and particularly in Africa.

It is very difficult to access accurate data on the extent of land appropriation in Africa. Government records are incomplete or do not translate from country to country, deals announced in the media via press conferences sometimes fall through, many deals may be unapproved and not appear on the government's radar, and many deals will be small deals that simply won't register. Nevertheless, Lorenzo Cotula and colleagues (ibid.) have attempted to gather data for selected African countries. Between 2004 and 2009, it is estimated that Ethiopia allocated 602,000 hectares to foreign investors, Ghana 452,000 hectares, Madagascar 803,000 hectares, Mali 163,000 hectares and Sudan 471,000 hectares.[18] The estimated level of investment related to the land deals in the sample of five countries is US$920 million, and the greater part of this investment is private-sector. To give these figures scale, a recent report in the *Observer* estimates that up to fifty million hectares of land – an area more than double the size of the UK – has been acquired in the past few years.[19] These figures, massive as they are, must be seen in country context. Some countries may have only limited access to land suitable for growing crops, a lack of access to water may also constrain production, and climate change projections generally indicate that African agriculture will be less productive in the future.

FDI in Africa almost doubled between 2005 and 2007 (UNCTAD 2008). This investment is uneven, however, with most being concentrated in countries with fossil-fuel and mineral resources – for example, Nigeria or Botswana. Other African countries risk being left behind, and their only natural resource of any value may be the land itself. Accordingly, governments are keen to encourage such investment, through a range of mechanisms such as direct land acquisition through government agencies, or support to the private sector in investor and host countries. Consequently, a diverse set of institutional arrangements can develop. In 2002, Sudan and Syria signed a Special Agricultural Investment Agreement which involves Sudan leasing land to Syria for food production for a period of fifty years (Cotula et al. 2009). Chinese state-controlled entities have been involved in the direct acquisition of land. Sinopec, one of China's nationally owned oil companies, is reportedly negotiating in Indonesia to set up biofuel plants and grow feedstocks there, with an initial investment of US$5 billion.[20] Private sector investor–state joint ventures are also popular. The most publicized deal, since scrapped, was that between the Madagascan government and the South Korean Daewoo Logistics company. This deal was to involve the acquisition of 1.3 million hectares of land to grow maize and oil palm mainly for export to South Korea.[21]

Strikingly, the most recent, largest land deals that involve the private sector have involved agribusiness and biofuels developers (ibid.). For example, Lonhro recently acquired 25,000 hectares of land in Angola, and is negotiating land deals in Mali and Malawi.[22] A Saudi consortium of agricultural companies recently announced plans to invest US$400 million in food production in Sudan and Ethiopia.[23] The UK energy company CAMS Group announced a lease of 45,000 hectares of land in Tanzania for investment in sorghum production for biofuels.[24] GEM Biofuels plc gained exclusive rights for fifty years over 425,000 hectares in southern Madagascar to plant jatropha for biofuel production.[25]

Investment in biofuels is driving changes in land ownership and land use for large tracts in Africa, whether directly through control of land to grow feedstocks, or indirectly to grow food to replace land lost to feedstock production elsewhere, or to hedge against rising food prices. In many respects this 'land grab' echoes old colonial relationships; rich, industrialized countries accessing natural resources in poorer, less developed countries. Thus we can witness Northern capital buying and leasing land across Africa. There are also new dynamics and relationships at play here; new actors, emergent economies or countries that had different names during the height of colonialism, are playing their part too in this scramble for Africa. Gulf States and richer East Asian countries are themselves accessing land in Africa, to compensate for their own investments in biofuel or to feed rapidly growing domestic markets. Land is fuelling new global relationships, and the matrix of subsidies, blending targets and interests means that biofuel-oriented land use is likely to tend to the large scale; this is likely to limit benefits for local people and the countries in which land is being appropriated (ibid.).

There is a long history of African countries, whether through design or pressure, focusing their agricultural production on export as opposed to meeting domestic needs. Debt, the price of oil and the thirst for FDI have driven African agriculture outside of its borders. This has fed into the globalization of agro-food systems more generally, the strengthening and lengthening of vertically integrated value chains, and power being systematically concentrated in the hands of fewer and fewer international players. Biofuels represent an opportunity for this agro-industrial complex to extend its reach into new commodities and new lands, and in doing so entrench global patterns of consumption and production. The small-scale biofuel producer, the small-scale farmer, the landless and, to a lesser extent, the African state itself risk being squeezed out, or down to the bottom, of the biofuel value chain. As we shall see in the next chapter, these topographies of power raise barriers to precisely the kinds of benefits that massive subsidy and global interest warrant.

Lusophone connections

We are now witnessing new sorts of relationship forming, quite different from the 'traditional' North–South or developed–developing-country relationships that were the norm. Brazil recently signed agricultural technical cooperation agreements in both Angola and Mozambique, which many hope will lead to the development of biofuels. The lusophone connection is perceived as a facilitator for these new relationships: 'The Portuguese-speaking synergy is likely to yield a ton of biofuel development.'[26]

Angola lost its sugar-cane plantations during its long civil war, but in 2009 a 30,000-hectare plantation was slated to open in Malanje Province. This plantation, with the sole purpose of producing sugar cane for bioethanol processing, represents the first major reintroduction of sugar cane in Angola in over thirty years. The plantation is a project jointly owned by Angolan state oil company Sonangol, Brazilian construction company Odebrecht and Angolan private company Damer. It is projected that the US$220 million programme will produce 280,000 tonnes of sugar, and 217 MW of power from burning the leftover sugar-cane bagasse.

Due east in Mozambique, similar initiatives are under way. In late 2009, Mozambique signed two accords with Brazil for a US$6 billion investment in biofuel production. An initial US$256 million has been invested in the Mozambique biofuels sector, covering 83,000 hectares. Some of the bioethanol produced from the sugar cane will be exported back to Brazil to supplement its own biofuels programme. While the Mozambican government's policy is that only 'marginal' or empty lands will be used for sugar-cane production and that all sugar cane will be refined in-country, there are reasons to be concerned. Mozambique's existing sugar-cane industry is not very competitive, not because it does not produce sugar cane cheaply but because the global price for sugar has been driven down artificially low, mainly by EU subsidies (Oxfam 2008). It remains to be seen whether the Mozambican government will maintain such an enlightened policy if sugar-cane-for-biofuels turns a bigger profit than sugar-cane-for-consumption. Such a diversion may of course be the correct thing to do if the sums add up, but as Chapter 3 illustrates, are we even sure what sums we should be calculating?

Reaching across the Indian Ocean

Like Brazil, India is also looking to Africa, driven by its need to reduce imports of crude oil and prompted by its slow-moving National Mission on Biodiesel. Facilitated by existing entrepreneurial networks, the rapid growth of its biofuel interests in Africa is impressive indeed. India has followed two main strategies with regard to biofuel investment in Africa. First, it has sought to utilize land for biofuel production, whether leased or purchased. Second, it has sought joint ventures, in particular investment in large, commercial biorefineries in Africa.

An example of the first strategy is that of Emami Biotech, part of the large Indian Emami Group, which has an agreement to plant up to forty thousand hectares of castor oil and jatropha plants in Oromia, central Ethiopia. This represents an investment of US$80 million. The hope is that this will produce up to 100,000 tonnes of biofuel feedstock.[27] This land was made available to the Emami

Group over a relatively short period of time, and Ethiopia has recently leased many large swathes of land, raising concerns about a possible lack of effective planning. It is not absolutely clear where the jatropha oil will end up. Some reports indicate the oil will be produced for consumption in Ethiopia and others that it is for import to India.[28] The latter is more likely.

The second strategy is evidenced in Mozambique, where Mozambique Biofuel Industries and two Indian partners, Malavalli Power Plant Private Limited and CVC Infrastructure India, invested €800 million to implement decentralized modular power units that can run off biofuels. The project draws upon Indian expertise in the development and implementation of relatively small-scale power units. In addition, Indian partners are providing management and finance for the development of the biopower units.[29]

Indian investment in biofuel development can be seen across Africa, not just on the eastern seaboard as might be expected. There are well over thirty large-scale biofuel partnerships between African and Indian interests in around a dozen countries. For example, biofuel production is being undertaken by a Ghanaian subsidiary of Hazel-Mercantile, a Mumbai-based distributor of chemical and petro-chemical products. India is the second-largest investor in Ghana, after China. The Indian government has provided US$250 million to the Economic Community of West African States (ECOWAS) for the establishment of a fund to promote biofuels in the West African sub-region. Tanzania can boast several new partnerships with Indian interests, focused mainly on jatropha production, with organizations such as Sumagro, Reliance and Biodiesel Technologies India. India has signed bioenergy cooperation agreements with several African countries, including recently Senegal, and Indian companies such as Emami and Naturol Biotech have signed cooperation and capacity-building agreements with African partners.

China is looking to follow India's lead; in many African countries China is now the largest overseas investor, and countries like Togo, Congo and Ghana are attractive places for Chinese biofuel investment, given China's huge reliance on crude oil imports. Within Africa, South Africa is investing in biofuel production and processing in its neighbours' and near-neighbours' territory, particularly

Mozambique and Angola. While looking at the ways in which biofuels forge new North–South relationships, we also have to be aware of the emergence of new South–South relationships, which represent emerging economies, their interests and their reliance on energy imports. Countries like India and Brazil are attracted to cheap, readily available tracts of land, and African countries are hungry for investment, expertise and access to technologies. There may be complementarities here, but there are significant risks, too, especially in the countries that choose to sell or lease land or find themselves in positions where they have to. These complementarities and associated exigencies are rapidly accelerating the rate of change, and the new relationships and partnerships that drive change need to be watched carefully if we are to dampen risk and amplify benefits.

Sustaining over-consumption

We are witnessing land-use changes of unprecedented scale, and new global connections of unparalleled scope. An estimated fifty million hectares of land have changed hands through hundreds of deals in Africa over the past few years.[30] It bears repeating that this is an area of land twice the size of the UK. Just five African countries have turned over an area of land to biofuel production slightly larger than Belgium (ActionAid 2010). If Europe is to meet a conservative 10 per cent biofuel blending target it would require a further estimated patch of arable land of 17.5 million hectares. Just acquiring this could push another 300 million people into food insecurity (ibid.).

The impact of a 10 per cent biofuel blending target, even assuming the most optimistic projections for GHG emissions savings, is negligible. In fact, as discussed in Chapter 3, if we were to factor land use into any equation, we might see no savings at all. Therefore, the *possibility* of GHG savings is an extremely expensive business. Taking into account the level of biofuel subsidy in the UK, and factoring in the most optimistic LCA, the cost of abating a tonne of CO_2 equivalent through the adoption of biofuels ranges from between €675 and 800, depending on the feedstock (Kutas et al. 2007).

Far larger GHG emissions savings could be achieved through simpler, cheaper sets of policies. For example, incentivizing or enforcing vehicle efficiency standards could accrue far greater savings than even second- (or third-) generation biofuels could ever hope to achieve. Current technology could reduce GHG emissions by as much as 30 per cent (HM Treasury 2008). Even if all biofuels consumed offered 100 per cent GHG savings (which they could not possibly), a 10 per cent biofuel blend would only be a third as effective on a per-vehicle basis. Attempts to implement policies of this nature have stalled, unfortunately.

Even simpler efficiency savings exist. Enforcing speed limits in the UK would reduce emissions by 8 per cent.[31] Subsidizing and incentivizing public transport use, promoting car-sharing, implementing congestion charges, adopting fuel-efficient tyres, driving smaller cars or riding a bicycle to work would all offer far more efficiencies, at far lower cost, and lower risk (Kill 2007).

One cannot help feeling, given the simple efficiencies that could be but are not being implemented through low-risk, low-impact technologies and policies, that neither reducing the EU's carbon footprint nor dealing with climate change are primary motivations. Actually expecting Europeans to alter their lifestyles or consumption patterns, even in ways that would imperceptibly alter their day-to-day routines, appears absolutely unconscionable.

Global biofuel assemblages

As this chapter has outlined, there are needs and priorities in the relationships between energy, environment and development that provide opportunities for innovative technologies. The beauty of biofuels is their ability to cut across problems or domains, and in effect provide a series of 'answers'. They are not necessarily the right answers, or the only answers, but they provide a framework within which policy and practice can mobilize.

From the perspective described above, the power of biofuels is not measured in kilowatts but rather in their ability to act as a conceptual pivot around which new ideas may be projected, knowledge generated and activities undertaken. The strength of biofuels – in policy terms far broader than other similarly potentially 'transformative' technologies such as biotechnology or nanotechnology – is their scope and their ability to appeal to so many interest groups – the state, the private sector, farmers, NGO workers and car manufacturers, to name but a few. They do not so much enable us to directly confront many of the most pressing global issues we face, as enable us to build alliances that satisfy ourselves that we are confronting those issues.

The evidence is clear that biofuels as policy ideas or solutions to present and future problems are gaining hold and momentum. A series of solutions to different types of problem (e.g. excess GHG emissions, energy security, rural underdevelopment), in entirely different contexts (e.g. the industrialized US Midwest, the emergent economy of Brazil, energy-starved Tanzania), are all mobilized around the same essential idea, that unlocking energy from biomass is efficient, clean and profitable; or at least more efficient, cleaner and more profitable. The beauty is in the ubiquity and universality, as those qualities render detail and context unimportant.

We can see, too, that biofuel assemblages take evolving forms and make new connections, at many different levels. Politically (or profitably) complex partnerships between agribusiness, farmers, NGOs and the state are forming, and these relationships are worked out in the equilibrium of profit, subsidy and sustainability that develops over time. In India, we can see that the promotion of jatropha planting and production is something that is being worked out through time. Issues are ironed out – or subsidized, key actors are drawn in. Ultimately, jatropha seeds will pour into refineries. The roles actors may play, and the roles of particular types of actors – for example, the state – constantly change, and that, too, speaks to the ability of the assemblage to shuffle actors and relationships.

Economically, the calculus of profit and/or subsidy triangulates where feedstocks are planted and harvested, in terms of how much energy is sold, and who benefits and who loses. The assemblage drives a recommodifying of nature, as previously 'unproductive' areas are now rendered profitable via jatropha seeds, new land uses promise more, and even newer land uses plug newly created gaps for crops and food. Future technologies promise yet more new biogeographies of profit, and this raises new sorts of promises (mainly for the developed world) and more problematically losses (mainly for the developing world). The attractive notion that biofuels can benefit the poor, rural and landless and somehow subvert what other types of agribusiness have summarily and consistently failed to do is another dynamic that shapes and is shaped by the global assemblage.

Environmentally, and perhaps most problematically, the potential of positive and negative effects in terms of GHG emissions shape policy, investment and well-being (even if only psychologically). We encourage land-use changes, some vast, that may alter, for good or for bad, our collective responsibility in dealing with climate change. The assemblage shapes scientific inquiry, funds it, prioritizes it, but perhaps does not want it to catch right up with practice. Land-use changes that can occur in a matter of days through mechanization can destroy ecosystems that have evolved for millennia, perhaps for ever. The direct and indirect consequences of these actions, and of biofuel production, are at best poorly understood, but to an extent that is rendered irrelevant, as we may need to invest in inefficient first-generation biofuels now, otherwise there may be no second generation. And a key driver of the assemblage is that there will always be another generation.

The real complication is of course that it is virtually impossible to pick the implications and interactions of biofuels apart from across and within the political, economic, environmental (and other) systems in which they exist and evolve. At stake are potentially irreversible changes being wrought in the fabric of our economies, societies and environments. Equally at stake is the threat that nothing is at stake. Poorer people may get poorer, but historically they do anyway; our economies will pursue other lines of growth, or will fall into recession, and politically there still, amazingly, appears little *real* will to undertake the kind of major structural changes needed to tackle climate change anyway.

Despite the uncertainty over impact, which tugs at the limits of our knowledge, there are some certainties, mainly moral ones. It is certain, predictable even, that we, via the biofuel assemblage, are imposing a kind of double- or perhaps even a treble-bind on the poorest and most vulnerable members of society. First, we are in effect asking the globe – humanity, everyone – to take collective responsibility for dealing with climate change. Recall the *World Development Report 2010* and its statement that the developed world should lead but the developing world ought to also do its part, and forgo the end days of the benefits of cheap fossil fuel. Second, we are transferring, in doing something or nothing, the risks of climate change, and of mitigation, on to the poorest people in the most vulnerable parts of the world. We are, in effect, expecting the rural poor in the developing world to alter their land-use patterns, their livelihoods and their externalities in order that we may maintain our consumption and energy-use patterns for as long as possible. Third, we are transferring the responsibility for inequity and for environmental care on to the most vulnerable, the groups of people and the countries that need justice and who generate less of an impact, as they always have, on the environment than the developed world has. This is profoundly unjust.

Calibrating change

Smolker et al. (2008: 38) note that 'the EU is reducing its own emissions by raising emissions in developing countries that produce the feedstock oils'. This risks turning the global South into a 'world energy farm' (McMichael 2009: 828). This in effect leads to a migration of sustainability. The biofuels assemblage acts as a planetary conduit through which trading emissions becomes a proxy for the trading of well-being and risk.

The assemblage coheres around the interests, and by Latourian extension the realities, of familiar and predictable sets of actors: the USA and the EU, agribusiness, petrochemical companies and the like. This coherence leads to the perpetuation of contexts which these actors seek: breathing life (quite literally, depending on one's Latin) into fossil fuels, allowing the EU to set and meet its emissions targets without needing to target the lifestyles of any EU citizens, securing a little, at least, more energy security for the USA and securing the profits and livelihoods of US agribusiness and farmers.

The assemblage deepens, and from alternative perspectives poisons or cleanses, long-standing North–South relationships (Dauvergne and Neville 2009). The assemblage also formulates new sorts of relationships – the relationship of Brazil with lusophone Africa, and Africa more generally. The means of food production become deformed and misshapen as fuel replaces food and international food supply replaces or supplements local food supply. The cheapest land, mainly in Africa, is purchased in order to replace food now converted to fuel. The circulation of globalization is given a jolt of bioenergy. The world, and nature and agriculture in particular, is being globally reconstituted, and a key element of this reconstitution is a recalibration of global risk. The 'migration of sustainability' is also a counter-migration of unsustainability (and irreversibility). Work-in-progress science, incomplete data and political policy reconstitute new types of risk in new places, in order to perpetuate old types of practice in existing places. Parts of Africa, already at risk of increasing food insecurity with more to come owing to climate change, change land use in order to allow Europeans to drive Land Rovers; millennia-old forests and grasslands are burnt to make way for a future that promises more of the same. There is also a subtext of a cumulative irreversibility of risk that we cannot yet comprehend.

This chapter has sketched the contours of a biofuel assemblage at a global level. It has identified some of the political drivers of biofuels as a new form of socio-technical system, and has highlighted some of the key convergences and interests, and the inherent logics that map out the topography of such an illogical system. The assemblage does not simply function to shape change at a global scale, it shapes new local dynamics, relationships and realities too. The next chapter will focus on the issue of locality, and, by extension, *scale*.

FIVE
Scale: solutions and risks

Global–local dialectics

We are witnessing the emergence of a globalizing, global biofuel assemblage. Powerful, global discourses of energy security, climate-change mitigation and development shape policies and subsidies that lever technological trajectories, new forms of land use and state priorities which are having profound effects on relationships between states. Within states, too, relationships are changing. How people earn a living, how they access energy, how governments believe they can best cater for the needs of their citizens are all being transformed in the face of the momentum of biofuels.

So far, we have traced the lineage of biofuel technologies, examined the propellants of a global biofuel assemblage, and delineated the limits of our knowledge of the implications of these technologies. The limits of our knowledge are not some abstraction; they are not, simply, an opportunity to reflect on how we perceive our trajectories of development, patterns of consumption and relationship to nature. These limits have the deepest implications for some of the poorest and most vulnerable people, communities and countries in the world.

This chapter will explore the implications of this and the points at which the assemblage is most firmly rooted, in the ground, within people's livelihoods, and in relation to our vulnerability and resilience to what goes on around us. In connecting up the local – to land grabs, land-use changes and new technologies, and ultimately to policy imperatives and global notions of how we should govern and shape the world – this chapter will show how the limits of the biofuel assemblage affect the limits of our ability to understand it, control it and foresee it.

Can biofuels be pro-poor?

There is a school of thought that biofuels can be pro-poor, supporting rural development and assisting developing countries with balance-of-trade deficits (UNDP 1995; Kammen et al. 2001). Regardless of the reality of this perspective, it segues into the pro-energy security, pro-growth, pro-technology perspective that drives the biofuel assemblage from the North. Both perspectives revolve around the globalization and liberalization of trade, and are built on a conception of development built around extending old and developing new forms of natural-resource exploitation. The huge growth in demand for biofuels, which has seen production double between 2002 and 2007 (UN-Energy 2007), may represent an opportunity for developing countries as tropical crops represent better returns in terms of biofuel yield per hectare than temperate crops. This, coupled with lower production costs, potentially represents a significant comparative advantage. Given the lack of clarity about whether biofuels are better for the environment than fossil fuels, might the potential of biofuels as tools of development provide an alternative rationale for investment, subsidy and trade?

First-generation biofuel processing technologies are widely available, relatively simple, and already in use in many developing countries. As an aside, one risk of the widespread adoption of second-generation biofuels might be that any developing-country comparative advantage is lost as production moves away from, for example, jatropha grown in developing countries towards the breaking down of already-produced agricultural wastes in developed countries. Regardless of whether first-generation technologies are used today or second-generation technologies are used tomorrow, the great attraction of biofuels is their similarity to petroleum-based fuels. They can be blended with petroleum fuels, they can be blended into fuel suitable for existing vehicles, they can be processed in simply modified existing facilities, and their transportation can make use of existing agribusiness value chains, which are already relatively well developed in most developing countries.

The pro-growth, pro-development perspective is based on the assumption that new, locally produced products diversify income, meet local needs and in doing so contribute to developing countries' balance of trade. While this may be true at the national level, and this is debatable based on experiences so far, what are the implications and developmental impacts when we focus down to the local and household level? Biofuel production generally requires extensive land and intensive labour. This means it has a direct impact on the communities where feedstocks are grown and processed. These impacts may be on land ownership, land use, labour demand, environmental change, investment and the local economy, and the resultant interactions of these factors may produce telling positive or negative impacts on different members of local societies.

The implications of land-use change for GHG emissions were discussed in Chapter 3; there are, however, other implications that may directly or indirectly affect the lives and livelihoods of developing-country peoples. Land-use change may dispossess rural people – generally poorer rural people – if they do not have ownership of land, diminishing their ability to access food. Some of the implications of this were discussed in the previous chapter. Land-use change may directly or indirectly influence food security, either by reducing the supply of food as land previously used for growing food crops is used for feedstocks, or by reducing access to food as prices rise as a consequence of production and demand (Pimentel 2004). Some of these impacts may be mitigated by promoting feedstocks that do not compete for land with food crops, such as jatropha, but if these crops prove sufficiently profitable they may of course be planted on land earmarked for food production anyway.

Large-scale biofuel production has similar environmental implications to those of other large-scale commercial agricultural production in terms of land degradation, water use and pollution, and the use of fertilizers, pesticides and other agro-chemicals. Many concerns are related to the manifestation of environmental issues that occurred during the Green Revolution, and many commentators are feeling a sense of déjà vu (Shiva 2009). Like large-scale agro-processing elsewhere, the production of biofuels generates waste products, which can have negative impacts on soil and water. A significant environmental issue more directly related to biofuels production is the loss of biodiversity through the creation of extensive monocultures that either replace existing land use or expand into new areas.

This is significant given that some of the prime regions for biofuels production also happen to be some of the world's most important biodiversity hot spots – tropical and subtropical rainforests and peatlands. The expansion of oil palm plantations at the expense of forest land in Indonesia and Malaysia, discussed in more detail later in this chapter, and the expansion of sugar-cane production into Brazil's central *cerrado* savannah and rainforest, are cases in point (WorldWatch Institute 2007). As demands for biofuels increase and endure, pressures on currently uncultivated land will amplify.

The effects of biofuels production on water are a further concern. Different crops in different locations require different amounts of water, itself a scarce resource in many of the regions where biofuels are grown. It takes 1,150 litres of irrigation water to produce 1 litre of bioethanol in Brazil, and over three times as much in India (de Fraiture et al. 2008). Of course, some feedstocks, such as jatropha, circumvent water demands somewhat by virtue of their drought tolerance. Water use for biofuels cultivation is no greater than for food crops, but as demand for energy and for food increases aggregate demand for water for agricultural production will also greatly increase. Water pollution is another concern, in terms of run-off from chemical inputs from feedstock production, especially contamination by pesticides, which can present health hazards for people downstream or involved in biofuels production, as well as having long-term environmental implications (Clancy 2008).

A final environmental consideration, which is also social and economic, is that of land classification. One of the advantages of certain feedstock plants, such as jatropha, is that they do not compete for land with food crops. The argument runs that biofuels feedstocks like jatropha can be planted on land that is 'degraded' or 'unused' (CFC 2007). Land that is 'unused' in a strictly agricultural or economic sense may provide important biodiversity to a region, or act as a wildlife corridor, provide access to water, or support other sorts of land-use activity such as the gathering of wild foodstuffs that may make important contributions to people's livelihoods (see Shackleton et al. 2007). Factoring implications and interactions hidden to policy into analysis is difficult but important.

Understanding social and cultural implications of biofuels production is also potentially complex, but important. Rural people are vulnerable to being displaced from their livelihood driven by the need to keep costs down through increasing mechanization of feedstock production on large plantations.[1] Mechanization, in turn, leads to the consolidation of land in order to squeeze maximum efficiencies from mechanization. The experience of the expansion of sugar-cane production in Brazil in the 1970s was of large plantation owners taking over small-scale farmers' land. In north-east Brazil in particular this led to social disruption and protest (WorldWatch Institute 2007). The expansion of oil palm plantations in Indonesia has led to similar unrest among poor farmers (Oxfam 2007).

Mechanization and consolidation may deny opportunities to sustain livelihoods from the cultivation of feedstocks, but other opportunities may be created within the agribusinesses created by demands for biofuels. Biofuels processing, for example, may provide alternative employment opportunities. Brazil is by far the most successful biofuels-producing developing or emerging country, and it has been claimed that over one million jobs have been created (Moreira 2006). Many of these jobs, however, may not be *new* ones. In São Paulo state, which is the most mechanized state in terms of converting sugar cane to bioethanol, the estimated employment generation is 2,200 jobs, with approximately another six hundred jobs in support (Kartha et al. 2005). Jobs are increasingly localized, so the impact on the wider economy is localized, and increasing mechanization tends to reduce the numbers of jobs while increasing pay and conditions for those that remain. The total employment claimed for Brazil's sugar-cane industry has declined from 670,000 in 1992 to 450,000 in 2003. Production doubled in that time period (WorldWatch Institute 2007). A further consideration, raised by Oxfam, is the quality of any jobs created. Jobs in less mechanized biofuels enterprises tend to be poorly paid, potentially hazardous, seasonal at best and day-to-day at worst (Oxfam 2008). In more mechanized industries, however, such as sugar-cane production in Brazil, wages tend to be higher than in other agricultural sectors, which may be attributed to more specialist skills being required, and may in turn be a driver of further mechanization (Kojima and Johnson 2005). There are always trade-offs.

It is important to recognize that many of the social implications of increased biofuels production are not inherent to biofuels production per se, although specific implications may be contingent on the nature of the biofuels crop and the local context.[2] Many of these dynamics are typical of processes of industrialization and expansion of agriculture in developing countries. Rather, as earlier discussions of Tanzania and land grabs, for example, have shown, investments in biofuels simply lay bare the power relations, vested interests and historical inequalities of the contexts in which they exist. Whether they amplify as well as underline these inequalities remains to be seen.

Indonesia, Malaysia and oil palm planting

From 2004, fuelled in part by the EU's commitments to blending of biofuels into petrol, the global price of palm oil rose almost three times from a historically steady US$380 per tonne to US$1,100 per tonne by mid-2008. Consequently, Indonesia and Malaysia committed 40 per cent of their total palm-oil output to biofuels (FoE Netherlands 2009). Their governments began to formulate policies and incentives, and the private sector began new investments in biofuel production (Kehati Foundation 2007). The aim of Indonesia and Malaysia is to meet 20 per cent of the EU's biodiesel demands directly through palm oil (Tauli-Corpuz and Tamang 2007).

In 2005/06, Indonesia also passed a presidential decree setting a domestic target of 5 per cent for the blending of biofuels into petrol by 2025, and establishing a national biofuel authority. Securing a substitute for oil is important for Indonesia as it currently imports and subsidizes oil to the tune of US$13.8 billion per annum – twice the amount of money it spends on education (Oxfam 2008). The hope is that making use of palm oil and reducing oil imports will save US$5–6 billion per annum, but this is not happening owing, ironically, to soaring palm-oil prices, which makes biodiesel uncompetitive in relation to petroleum products.[3]

Prior to this nascent biofuels boom, only 3 per cent of the land area of Ketapang, part of West Kalimantan province in Indonesia, was devoted to oil palm plantations. This amounted to approximately 100,000 hectares, which by the end of 2005 had risen to 742,000 hectares; by 2007 over 1.4 million hectares had been allocated to palm-oil production. This represents over 40 per cent of the district's total land area. Ketapang was not unique; growth of this magnitude occurred across Kalimantan (FoE Netherlands 2009).[4]

The sharp global price rises for palm oil, coupled with the stimulus of government policy, have driven deforestation and replacement with oil palm plantations in an unregulated, unplanned way (ibid.). Strict guidelines in Indonesia regarding the use of Environmental Impact Assessments and government regulation of new plantations via licensing and certification have had little impact in terms of managing plantation development. A Friends of the Earth report identifies 'fast track' licensing, with de facto waiving of legal requirements designed to protect the environment and local communities, and secure state income. Many companies, including subsidiaries of multinational agribusinesses such as Cargill, have commenced plantation development without having secured the necessary approvals (ibid.).[5]

The Ketapang district government has issued several dozen oil palm permits for land that either completely or partially overlaps with 400,000 hectares of protected forest land. This protected forest land is the site of unique biodiversity and provides a home for many species under threat of extinction, most visibly the orang-utan.[6] If encroachment of oil palm plantations on protected forest land continues at present rates the orang-utan may become extinct in seven years' time (FoE 2005).

Palm-oil prices should in theory be of benefit to poor producers, although rising international prices do not translate perfectly to rising farm-gate prices (Oxfam 2008). Intuitively, one would expect benefits for other people involved in palm-oil production and knock-on benefits for people living in close proximity to emerging local economies. Indeed, Indonesia's palm-oil development has been touted by the government as a means of poverty alleviation. The case of Indonesia, however, highlights several issues in this regard. First, concessions given to commercial plantation operations often overlap with indigenous peoples' customary land, giving cause for conflict. Land has been developed without discussion with or the consent of local communities, and plantation companies have failed to develop legally required local development projects (FoE Netherlands 2009). Second, plantation workers tend to be underpaid and employed on a day-by-day basis. Children are commonly employed. Workers are periodically exposed to hazardous working conditions (Wakker 2004). Third, and finally, smallholder palm-oil producers risk being squeezed out. Government regulations require commercial plantations to develop at least 20 per cent of their land into community projects, but there is little evidence of this happening in Ketapang, at least (FoE Netherlands 2009). In addition, smallholders are subject to unfavourable terms as they typically depend on the commercial plantations for access to processing facilities. Power is skewed towards the large, commercial concerns at every turn (Peskett et al. 2007).

Indonesia's experience epitomizes the crux of the food-versus-fuel debate. Palm-based cooking oil is a staple of the Indonesian diet. Driven up by the global price of palm oil, the consumer price of cooking oil went up 40 per cent to 2007 (and by 70 per cent in Malaysia).[7] Palm oil, of course, does not have jatropha's benefit of being non-edible and so-called 'edible oils' have witnessed the highest price rises of any foodstuff over the past few years. This is no trifle. Cooking oil is a considerable expense for poor South-East Asian households, and of course the poorest households suffer the most.

Shortages, queues and sporadic outbreaks of violence have occurred across Indonesia and South-East Asia, and Oxfam reports that food vendors and home industries have been forced out of business as they can no longer produce oil-based food products cheaply enough (Oxfam 2007). The Indonesian government was obliged to raise export taxes on palm oil and reduce import taxes on substitute cooking oils.

Both Indonesia and Malaysia, in keeping with their 40 per cent biofuel commitment, have invested heavily in processing facilities. By 2007, Indonesia's production capacity topped 2 million tonnes per annum (Oxfam 2008). One year later, seventeen facilities had suspended operations and only five facilities remained in production at reduced capacity.[8] The international rises in the price of palm oil meant that domestic biodiesel producers simply could not afford to purchase palm oil.[9]

The experience of Indonesia and Malaysia is enlightening. Extending their gaze internationally and seeking to take advantage of (partially) biofuel-driven palm-oil price rises have been both disappointing and problematic. Prices have risen beyond what domestic biodiesel producers can pay, negating, for the moment at least, considerable investment. In tandem, edible-oil prices have been driven up internationally, which has had huge implications for poor households in countries where palm oil is an important part of the diet.

Indonesia and Malaysia find themselves caught in a double bind of missed opportunity and household pressure, driven by EU policy. This underlines the fact that, despite thoughtful policies, investment and a rational response to the market, even if the market is distorted by subsidy, the short- and medium-term developmental implications of investing in biodiesel production and risking the vagaries of the market may be damaging. And that is before one even considers the longer-term environmental implications of rendering species extinct, destroying millennia-old rainforest and replacing it with oil palm plantations.

Palm oil represents something of the tension of scale. Large-scale, often internationally owned commercial plantations dominate and squeeze out local smallholders. At the same time the Indonesian government orients policy towards high international palm-oil prices but becomes a victim of the international market. Complex relationships and feedbacks exist between the international context – where demand for biofuels manifests itself – and the local context – where crops are grown. Where these relationships are driven by multinational agribusiness and are mediated by volatile commodities markets the state risks becoming irrelevant and powerless to secure its own food, fuel and development opportunities.

Malian small-scale solutions?

A logical way to avoid the vagaries of market and the whims of Northern policy is to focus only on national and local production. Examples of emerging small-scale projects do exist, and it is worth reflecting on their potential benefits and the constraints they face. It is easy to forget or assume they don't exist given the huge emphasis on using plants to power vehicles in developed and emerging economies.

Mali has a longer history of working with jatropha as a potential feedstock than most other countries. Several donor-funded projects have been developing jatropha energy projects for over twenty years. In 1987, the Special Energy Programme of the Ministry of Cooperation funded a project that eventually lasted for ten years. Since 1993, UNDP/IFAD have supported a project in Mali and Burkina Faso that has developed 'multi-functional energy platforms' which use jatropha-derived fuel to provide energy for light, electricity and the pumping of water (UNDP 2004). GTZ have also funded research into jatropha in Mali. Since 1999, Denmark has funded the 'Nordic Folkcentre for Renewable Energy', which has experimented in jatropha cultivation for energy production.[10]

In some respects Mali is an obvious site for all this attention. Ninety-nine per cent of the rural population lack access to modern energy services, and it is clear that the state and parastatals do not have the capacity to provide the services needed given the diminishing value of Malian exports (mainly cotton) and the rising price of crude oil (FAO 2009b). Jatropha itself also has a history in Mali. It is believed it was brought to Africa by Portuguese explorers and it has been used for decades as a natural means to stop animals eating crops and as a guard against soil erosion, its natural toxicity meaning that it is less attractive to pests (Jongschaap et al. 2007). There are an estimated 22,000 kilometres of jatropha 'fence' in Mali.[11]

The Mali Folkcentre, funded by several organizations, including a government parastatal, initiated a new project in 2006 in Garolo commune, to pilot jatropha as a means of providing electricity in small communities. This initiative was in part precipitated by the increase in crude oil prices in 2005, which led to strong political support from the government.

The project promoted jatropha cultivation in relation to existing farming practices. Intercropping was encouraged, as was making use of jatropha residues as fertilizer and as a binding agent to counter soil erosion. This approach built on Mali's existing experience in growing jatropha and minimized emissions and other environmental impacts, while intercropping mitigated food insecurity.

Alongside cultivation, a jatropha value chain was developed in partnership between the Garolo Jatropha Producers' Cooperative and a local private sector power company. The cooperative deals with all issues regarding the cultivation of jatropha and the production and sale of jatropha oil and other residues. The power company is responsible for the generation and sale of electricity. Currently, electricity is generated and supplied to around 250 households (FAO 2009b).

The entire undertaking has been developed around small-scale farmers. Jatropha cultivation is growing rapidly, owing partly to initial guaranteed prices to fuel the Garolo power plant but also in response to other demands, from entrepreneurs and foreign companies willing to purchase and process the seeds to produce biofuels for national and international markets. The Malian land tenure system and customary law potentially raise obstacles as it is considered that land planted with trees definitively belongs to the person who planted them. This is a constraint for farmers who are considered to have the right to cultivate for life, but are not fully fledged landowners in their own right. Part of the role of the cooperative is to deal with issues such as this.

In a region with relatively few opportunities for cash generation, the opportunity to earn income from selling jatropha seeds and processing them into oil is a welcome opportunity. Consequently, there is much interest among local farmers in getting involved in the cooperative. There is a discernible diversification away from cotton plantations, which is a good livelihood outcome and places less stress on the environment, as mature jatropha plantations do not require irrigation. The small-scale, dencentralized approach means that the entire value chain exists in and around Garalo, which helps to build trust and social networks around cultivation and processing. The local approach helps limit the impact of any drop in global crude oil prices, which might lessen international demand for jatropha. Finally, there are also indirect benefits – new micro-enterprises have sprung up thanks to access to electricity, and if the approach is scaled up and replicated across many other villages it has the potential to reduce the Malian government's reliance on fossil-fuel imports.

The Mali Folkcentre pilot is a substantial and complex under-taking, and government support – political and financial – has been crucial. Alternative energy generation, focused specifically on jatropha, is part of Malian government policy.[12]

The power plant has been developed to run on both pure jatropha oil and diesel. By 2009, 5 per cent of the fuel for the plant was pure jatropha oil, and it is envisaged this will increase rapidly to reach 100 per cent by 2013. So far only 600 hectares of jatropha have been planted by around 250 small-scale farmers: a small percentage of the ultimate target of 10,000 hectares. Nevertheless, electricity is already being produced and careful planning means that the conditions necessary for a complete transition to using jatropha oil for electricity generation seem to be in place (ibid.). The approach may provide a model for replication across rural Mali, and perhaps elsewhere in Africa (van Eijck and Romijn 2008).

Kenya – striking a balance

Kenya has a history as an early adopter of technology in Africa. The country spends more on agricultural research than any other sub-Saharan African country bar South Africa. Several international agricultural research centres are based there. It hosts the African headquarters of the United Nations, which brings with it the head-quarters of several international non-governmental organizations and a constellation of regional and local-issue-focused organiza-tions. It is little surprise, therefore, that both the Kenyan government and local entrepreneurs have been exploring the potential of biofuels.

On the face of it, Kenya's policy seems ordered. Since 2004, government policy papers have explicitly recognized the need to encourage the wider adoption of renewable technologies, especially biodiesel (Mouk et al. 2010). In 2006, an Energy Act mandated the government to pursue and facilitate the production of biofuels, with-out specifying how this would be achieved. While the Ministry of Energy has been leading on biofuel development, other government departments are mandated to play a role, including the ministries of Agriculture and of Forestry. Other active ministries include Trade, Industry, Water Development, Transport, and Environment and Natural Resources.

Beneath, within or between ministries, a bewildering array of government commissions and bureaus have responsibilities and remits that intersect with the development of a biofuel industry. The Energy Regulatory Commission is responsible for regulating the energy sector. The Kenyan Bureau of Standards, however, does not make it clear where it is permissible to produce, sell or use biofuels. One interpretation might be that small-scale production and use is probably acceptable, but the Bureau of Standards ought to adopt relevant standards before large-scale commercial biofuels activities begin. Other policy gaps exist – while there are standards for blending up to 10 per cent bioethanol into petrol as long as the resulting fuel blend meets existing standards for petrol, there is no equivalent provision for biodiesel. Given that jatropha production is being pursued on a number of fronts, this is a significant omission.

A number of other government entities are responsible for aspects of biofuel production and use. The National Environmental Management Authority is mandated to create incentives for the promotion of renewable sources of energy. Through the Environmental Managements Coordination Act, environmental impact assessments must be undertaken in relation to any biofuels standards that are proposed.

The Ministry of Energy has attempted to coordinate the plethora of ministries, departments, commissions and acts through the creation of a National Biofuels Committee, which brings together public sector, private sector and NGOs that participate in the Kenyan bioenergy value chain.

At the moment, however, there is a policy and a strategy gap. This gap, coupled with Kenya's preponderance of NGOs, its research capacity and its relatively capable local private sector, means that uncoordinated activity is a real possibility. This is a particular concern for developing countries which has been raised by people working on comparative biofuel policy. For example, in 2008 a Kenyan court temporarily halted a US$370 million sugar and biofuel public–private project in a coastal wetland following warnings from conservation groups that livelihoods and wildlife would be threatened.[13] Kenya's National Environmental Management Authority originally cleared the project, which would have produced 23 million tonnes of bioethanol per annum, but complaints have led to a judicial review. Circumstances of this nature indicate two things – the difficulties

Kenya faces in coordinating across so many entities with partial responsibility for biofuel development, and the lure and the risk of significant private sector investment.

Kenya's draft bioethanol and biodiesel strategies identify the opportunities that biofuel development can bring – income to rural areas, easing of rural unemployment, the environmental benefits of blending biofuel into conventional fuels, and the diversification of energy sources (Ministry of Energy 2008). There is recognition, however, that while the focus of Kenya's strategy should be on assisting smallholders to produce jatropha there are serious deficiencies in the jatropha value chain that need to be addressed through private sector participation. In addition, while many of the areas of the country identified as suitable for jatropha production are arid or semi-arid, and it is therefore unlikely to compete with traditional food crops, there is sufficient ambiguity to ensure that encroachment into more fertile, productive areas is not ruled out. This raises two issues. First, there are productive livelihoods in arid and semi-arid areas; they are just not recognized as such through the rubric of agriculture. Emerging research shows, for example, that the semi-nomadic, pastoralist Orma of south-eastern Kenya are finding that their grazing lands are being squeezed as land is earmarked for feedstock production.[14] Impacts on transient livelihoods appear simply to be overlooked in Kenya, as elsewhere. Second, research is increasingly showing that jatropha grows much better on fertile land, and that there is a yield threshold below which it is simply unproductive to cultivate it given the relative labour intensiveness of the process. Thus, while Kenya's policy might be to focus production elsewhere, the reality might end up being quite different, particularly given the acknowledged reliance on private sector investment and activity necessary to ensure functioning value chains (Vianello 2009).

A further troublesome element of Kenya's biofuel engagement is the stated aim of attracting significant international investment to stimulate the industry. It is unlikely that the Kenyan government would be willing to lose such investment on the basis of restricting jatropha cultivation to areas of the country that would not compete with food production. Kenya's cut-flower sector would be a case in point.

Along with uncertainty about the role of the private sector there is a vagueness of policy and a lack of clarity over which part of the government is responsible for precisely what. Kenya's situation may be better planned than that of neighbouring Tanzania, for example, but it has its own pressures and context to deal with.

In a similar vein, there are many actors who work in murky spaces between the non-governmental and private sectors. Organizations such as the Vanilla Development Foundation and the Green Africa Foundation are exceptionally active in promoting jatropha and other oilseed plants as viable opportunities for rural development and investment. These organizations may play an important network-ing role, connecting up disparate actors, but they are invariably operating in an environment of institutional confusion and limited organizational capacity. Kenyan civil society is dynamic and eager to seize new opportunities, and this is a strength, although it is not always clear whose interests these organizations are promoting – those of the rural poor, local investors or overseas stakeholders.

While Kenya is working out policy and assigning responsibility for various aspects of its nascent biofuel industry there remain sig-nificant gaps and uncertainties: the balance between local needs and development opportunities and foreign investment and interests, the coordination of activities across the multitude of state and non-state actors, the future balance and prioritization of activities related to biofuels, and ultimately the shifting international context and its demands for biofuels, whether first-generation or future-generation. There is a pressing need for scale and balance in policy formulation and implementation. As we have seen, this may be a local or a national need, but it should also be an international responsibility.

Scale and perspective

This chapter has identified threats for poor people and poor coun-tries. Opportunities do exist, however, embedded in these threats. Brazil has shown that is it possible to incubate a successful biofuels sector – although within that sector benefits and risks accrue dif-ferentially to those with resources and those without (Wilkinson and Herrera 2008). The early stages of the Brazilian programme have been described as having increased the concentration of capital, land

and power, and of commodifying rural labour.[15] Brazil's bioethanol sector, despite policy and purported pro-poor emphasis, has driven greater inequality in its extension of the reach of capital into the countryside.

The cases of Indonesia, Malaysia and Kenya, too, highlight the contestations and risks that poor people face when confronted by the advancing frontier of biofuels development. The global assemblage of biofuels links new locales into global value chains that are driven by the power of capital and politics. Biofuels connect people up in new ways, and this inevitably exposes some people to new profits and some people to new risks. These disparities are calibrated by people's access to resources and ability to exert power over these relationships. Policies may seek to balance out these disparities but experience of the globalization of agro-food systems suggests that inequality will always wriggle from the grasp of policy, no matter how well thought out (see McCullough et al. 2008). Biofuels represent both new formations of global value chains and all that infers for producers and promises for actors at the peak of the chain, and new frontiers of the relationship between capital and nature (McMichael 2009).

Local implications seem to get lost in the distinctive global nature of biofuels, with their ability to intersect with global issues such as climate change, energy provision, development and maintaining economic growth. The primacy of global biofuels development risks skewing our perspective of what is sustainable, and indeed what sustainability is. The framing of biofuels as a solution to global problems generates policies that reinforce this: for example, Indonesia and Malaysia dedicating 40 per cent of their palm-oil production to biofuels production for international markets; the EU securing biofuels – such as palm oil – from elsewhere as the only hope of meeting its blending targets; Brazil seeking international markets, having developed a strong internal bioethanol production system; the USA processing 25 per cent of its corn crop into biofuels. Investment and profit push the expansion of biofuel production systems. This expansion is also fuelled by policy and subsidy, but is increasingly being driven by large agribusiness. Companies like Archer Daniels Midland and Cargill have invested heavily in US bioethanol production, and are moving into international markets, looking towards places like Indonesia to develop plantations. We can see, in turn, other actors orienting themselves towards global policy perspectives as clear incentives encourage them to do so. Policy is often framed around other priorities. Subsidizing biofuels production supports US farming interests and sustains the oil industry by imperceptibly 'greening' it. It also justifies current levels of energy consumption by proposing and providing a partial solution. There is little emphasis on generating efficiencies, only on generating new sources of energy. The relevance of scale is obvious: per capita oil consumption in the USA is over one hundred times that of Tanzania, for example (Oxfam 2008). Policy seeks to sustain the very practices that led to unsustainability.

Focusing on the globalization of solution and the sustainability of consumption focuses solutions on large-scale, export-oriented biofuels. Focusing on innovations in local, small-scale, contextualized biofuels systems, such as in the case of Mali described earlier, is rare. The role of biofuels as a means of development in developing countries is clearly a second-tier aim in relation to developed-country energy security. Small-scale solutions and other options, such as focusing on directly deriving energy from biomass, remain on the periphery.

When small-scale solutions for local needs are noted they risk being transformed into large-scale solutions for global issues. The lure of profit may ramp up small-scale solutions. Crops such as jatropha that can grow on marginal or non-arable land in order to avoid food-versus-fuel production conflicts may end up being grown on productive farmland if sufficient profit can be made. Alternatively, the prospect of enhanced profits may mean low-input crops such as jatropha may be more intensively managed in order to increase profits, negating GHG emissions savings. In this way collective benefits, such as GHG savings, are prioritized below individual benefits, such as enhanced profits.

The global biofuels assemblage depends upon and drives dispossession and land-use change. Globally, land use is shifted from local food production to industrial purposes such as the cultivation of feedstock for fuel production, from small-scale production to plantations, and from local needs to new global priorities. These changes, dramatic as they are, are not in themselves about change; they are about maintenance – maintenance of elites, maintenance of interests, maintenance of patterns of production and consumption, and maintenance of power. Essentially the future will maintain the past.

Scale is temporal, too. The transition from so-called first- to second-generation biofuels generates new risks and closes existing opportunities. Second-generation technologies do not depend on feedstocks like palm oil, jatropha and sugar that are already well established in tropical areas. Second-generation technologies depend on ligno-cellulosic materials, which require advanced, developed-country processing techniques. The risk for developing countries is that developed countries can use second-generation technologies to unlock energy from feedstocks that grow well in temperate conditions or from existing agricultural wastes. This will negate any competitive advantage that tropical, developing countries might currently enjoy.

There are two ironies in this. First, one of the rationales for investing in first-generation biofuels technologies, despite their so far dubious GHG emissions savings, is that we need to go through this generation of technologies in order to develop new, more efficient techniques. Thus, developing countries risk becoming ecosystem-sized laboratories for new technologies that will ultimately benefit developed countries. A second irony is that another of the rationales for developing second-generation technologies is as a direct response to the first-generation food-versus-fuel dilemma. Perhaps greater caution needs to be exercised for fear we look to the future without considering the day-to-day implications of biofuels production that people, communities and countries currently face.

The global reach of biofuels

Biofuels are inherently neither pro- nor anti-poor. They do, however, represent a new global compact between the rich and the poor. The poor are being asked to take responsibility and compensate for the political, energy and environmental needs of the rich. This compact is veiled within value chains, policies and new technologies, but it is there. Hidden within the interaction between these relationships, policies and technologies are significant risks for the communities in which feedstocks are being produced, and for people who perpetually live on the edge of food security.

It is difficult to attribute specific impacts to specific biofuels initiatives. There is a lot of evidence already, even though we are still in the early stages of global investment in biofuels, that significant social, economic and environmental impacts are being generated. From a global perspective, in which biofuels are meant to make important contributions to lowering GHG emissions, these impacts are largely invisible. From a Northern perspective – from a petrol station forecourt where we can fill our car, satisfied that we are doing our bit for the environment by pumping petrol blended with biofuel – these impacts are also invisible.

Perhaps these impacts are costs worth paying if they help us manage climate change? Perhaps these costs are outweighed by the benefits of national energy security? Perhaps we should accept negative impacts for some if others reap positive benefits? The fundamental issue with these questions is not the questions themselves (unpalatable as they may be) but the issue of who is answering them. Questions of this significance ought to be carefully and collectively considered, not obscured by a veil of exigency, profit and opportunity. The concluding chapter will reflect on some of these issues.

SIX

Sustainability? The globalization of risk

Scale and scope

During May 2010, Craig Venter and his team announced the creation of the first 'synthetic life'. *Mycoplasma mycoides JCVI-syn1.0* was created by synthesizing a DNA molecule containing an entire bacterium genome, and introducing this into another cell (Gibson et al. 2010). There has been debate, scientific and moral, about whether *Mycoplasma mycoides JCVI-syn1.0* indeed constituted human-created life. There has been much less debate about the ultimate purpose of this line of research. Venter's research takes place in an eponymous research institute, is backed by venture capital, and profitable application is the ultimate aim. In a 2007 interview, when asked what one would do with synthetic bacteria if one could create them, Venter replied:

> Over the next 20 years, synthetic genomics is going to become the standard for making anything. The chemical industry will depend on it. Hopefully, a large part of the energy industry will depend on it. We really need to find an alternative to taking carbon out of the ground, burning it, and putting it into the atmosphere. This is the single biggest contribution I could make. (Aldous 2007: 57)

Venter is serious. He began talking about the potential of synthetic bacteria generating energy several years ago. In 2009, he entered into a US$600 million partnership with Exxon Mobil to create biofuels from algae.[1] His vision is to build 'an entire algal genome so we can vary the 50 to 60 different parameters for algal growth to make superconductive organisms'.[2] Venter has coined the concept of 'fourth-generation' biofuels, a hypothetical domain of genetically optimized feedstocks and fuel-producing genomically synthesized microbes. Venter's is indeed a brave new microscopic world.

It is telling that the notion of biofuels now cuts across the consciousness and scope of so much of science, present and hypothetical; microscopic solutions for macroscopic problems, genomics for global concerns. A platform technology is a technique or tool that enables a range of present and future scientific investigations – for example, bioinformatics enables a range of data assembly and analysis that would not have been possible beforehand. Biofuels have virtually become the antithesis of a platform technology; they provide scope for multiple streams of research.

The scope of biofuels now extends from the possibilities of the descendants of *Mycoplasma mycoides JCVI-syn1.0* to the most complex of global value chains. Superficially at least, one might argue that a shared notion of sustainability, of a more secure, collective future, is the connective tissue for all this endeavour. On the contrary, I argue that risk, not sustainability, generates the connections in a biofuelled world. As previous chapters have shown, a complex web of interests, politics, science and consumption has driven the biofuels agenda. These constituents coalesce to do two things. First, they generate new, more profound risks (and responsibilities) for certain populations in certain parts of the world – the rural poor, the hungry and the vulnerable. Second, they perpetuate broader risks for all of us in enabling us to sidestep the profound issues of consumption, over- and underdevelopment, and difficult choices that we surely must face. Somewhat ironically, invoking biofuels in the name of sustainability allows us to completely disregard what sustainability actually means. That is the most profound risk.

Causes and effects

Perrow's *Normal Accidents* (1999) argued that complex interactions (particularly manifested in risks that are unknowable, unobservable, new or delayed in their manifestation) combined with tight coupling (closely connected, dependent events) would generate technological systems that almost inevitably fail through their sheer complexity. Furthermore, typical institutional precautions and responses to failure generate further complexities, and unknowns may generate new opportunities for failure. In sum, failure is almost hardwired into complex technological systems, and most attempts to mitigate failure amplify a most profound risk of failure, hence 'normal accidents'.

The biofuel global assemblage, in its current configuration and through its current drivers, represents an exemplar of a complex, tightly coupled system. In fact, it represents possibly the most complex configuration of systems shaped by a modern technology. The feedbacks, interactions and impacts are multilevel, multivariate and unprecedented in their implications.

For every policy assumption there is an experiential counterpoint. For every tale of success, there is a vignette of failure. For every positive, there is most probably a negative, somewhere, even if we do not recognize it yet. This in itself underlines the limits of our knowledge and our inability to shape and control the systems that biofuels interact with. To list but a few.

An initial assumption of land-use policy is that feedstock production need not compete with food production as new feedstock crops such as jatropha can grow in areas unsuitable or unproductive for food crops. Global rises in the prices of staple food suggest that a much more complex relationship exists between foodstuff and feedstock than we are willing to admit (Mitchell 2008). Experience in India is also beginning to show that low-yield jatropha production is inefficient and production on less marginal land is necessary to incentivize investment. This requires new mechanisms to manage land use to ensure that food crop production is maintained. Similarly, low yields are themselves (contradictorily) assumed to limit competition for land use. If yields are increased, more productive crops are planted or energy is more efficiently extracted, competition is diminished. In contrast, history and experience show that demand for land is driven by intensification and increasing yields act as a stimulus to further intensification and commercialization. Generating greater efficiencies simply increases the financial incentives for a shift to agro-industrial and monoculture systems. As earlier chapters illustrate, we can witness this happening in South-East Asia, for example.

An extended assumption is that greater production can alleviate competition among diverse uses of biomass – for example, by limiting food-versus-fuel competition by ensuring the technical ability to meet demands for both markets. The reality is that biofuel crop expansion responds to global markets (like all internationally traded agricultural commodities), where high prices and greater demands readily consume any extra production or yield, especially given the global linkage between feed and fuel markets (Oxfam 2008). This is especially a problem for commodities such as jatropha seeds, where there is no possibility of creating niches for producers (by local specialization through avenues such as geographic indicators, for example), the seeds are relatively easily transported, and policy is ensuring that demand is increasing in many places. This even endangers success stories, such as Mali, if it becomes more profitable to export jatropha than to use it to produce energy locally, and of course further negates whatever marginal environmental benefit may exist.

New technologies are assumed to generate efficiencies in energy production, through practices such as more efficient use of agricultural residues or the combination of other technological innovations, such as agricultural biotechnologies (Royal Society 2008). In reality, new efficiencies are likely to generate further economic incentives for monocultural systems to supply biomass to centralized biorefineries. We can witness this in the growth of commercial palm-oil ventures in Indonesia, for example. This may enhance one output of the system, but risks negating any benefit for small-scale producers. Other assumptions are built into thinking about efficiencies. Agricultural wastes have other, current uses – fodder, soil cover, fertilizer, for example – which means there is an explicit trade-off in using them in biofuel production. 'Waste' is rarely waste, much as 'unproductive' land is rarely entirely unproductive. Indeed, past experience shows that focusing on planting marginal land as a means to avoid conflict with food production has repercussions. The example of Kenya shows that using semi-arid land for agricultural production has badly affected pastoralists. Indeed, there is a long history of agro-industrial production extending into common grazing areas (McMichael 2009). Land is defined as 'marginal', meaning that it has not added value to global markets, rather than that it has no utility at all (Meillassoux 1981).

A central assumption is that biofuels offer the possibility of new livelihoods and income streams for rural producers and developing countries. New technologies create new markets and generate new income streams. On the contrary, evidence thus far suggests that while small-scale rural producers may derive benefits in certain situations and contexts, it is mainly larger-scale commercial producers who benefit (Oxfam 2008; Christian Aid 2009). This, in turn, puts additional pressure on small-scale producers' access to land and may risk limiting their income streams. Additionally, local employment on commercial farms is limited, poorly paid and not secure. Finally, value addition in processing biofuels is centralized and generally requires heavy investment, which means incomes and opportunities are further limited. This, of course, should come as no surprise as it is reflects much of the history of the evolution of global agro-food chains (see Patel 2007).

A final assumption is that new, latter-generation biofuels will solve virtually all of the problems and contestations raised by current first-generation biofuels. If we assume the science will work and that second- and later-generation biofuels do become commercially viable (and that is a big if) this still leaves the huge issue that developing, tropical countries lose their comparative advantage. Instead, we will have encouraged investment and production in first-generation biofuels in regions of the world that will be ill equipped to reconfigure their production towards other agricultural commodities. Essentially, subsidy-driven supply and demand will have locked countries into unsustainable practices which they can ill afford to extricate themselves from. Scientific cornucopianism might be alluring, but it certainly won't be beneficial for all.

This section has not attempted to provide an exhaustive list of assumption versus reality, or causes and effects; that would be far beyond the scope of this book, and indeed probably any other at this stage. To quote Andrew Moore: 'As far as biofuel production is concerned, we are still in the Stone Age' (Moore 2008b: 99). Rather, this section has simply highlighted a few of the multiple disconnects between assumption and practice, and illustrates again the need to learn from our history, and in particular the implications of promoting global agro-food (or feedstock) value chains. At the moment we do not seem to have more than an inkling of the systemic complexity of the systems we are busily creating.

Assembling biofuels

Biofuels represent both an opportunity and a risk. As we have seen, we risk biofuels binding us to irreversible decisions regarding land use, investment and how we articulate our relationship to nature. They do also present opportunities, perhaps to produce fewer GHGs, perhaps to provide energy sources for communities that need it, perhaps to earn, or save on, foreign capital. There may be tangible benefits and tangible impacts.

Possibly the biggest opportunity that biofuels present to us is conceptual, however. Perhaps biofuels present us with the insight to rethink many of our relationships. We need to rethink our relationship to nature, to the economy (or the market), and to each other. Biofuels are currently being proposed as a universal techno-solution that will *not* require us to rethink the connections that shape our world. In fact their uniqueness lies not in their ability to unlock the power of sunlight but in their capacity to absorb change and absolve us from effecting it. The biofuel assemblage apparently offers us a system to make sense of a world composed of competing demands, future uncertainties and difficult decisions that urgently need to be made. The assemblage stabilizes these uncertainties and presents simple solutions with which we can navigate complexity.

The biofuel assemblage is beguiling. It generates certainties, from some perspectives at least, provides solutions and highlights pathways to the future. It stitches together techno-scientific systems, political imperatives and economic rationalities, and in doing so stabilizes one vision of sustainability. We would be wrong to assume that the biofuel assemblage, the global apparatus that propels biofuel investment and activity, is monolithic and insurmountable. It is in a constant state of flux. New uncertainties and emerging critiques are addressed and absorbed into the logics of the system. Critiques of first-generation biofuels are not taken as critiques of biofuels per se; they become rationales for second-generation biofuels, and so on. Calls for more research into the environmental implications of biofuel production are not ignored, as new research can be framed in such a way as to provide fresh impetus to the assemblage. Under-investment must be met with more investment. Negative impact must be met with better policy and instruments of governance.

Bruno Latour (1996) talks about new technologies existing only insofar as they can generate networks of support. These networks are generated by the promises of technological projects, and framed by the experts and interests that instigate, map and translate them: 'By definition, a technological project is a fiction, since at the outset it does not exist, and there is no way it can exist yet as it is in the project phase' (ibid.: 23). At one level, the biofuels assemblage creates future fictions out of contemporary realities. The power of biofuels is not so much in their capacity to generate sustainabilities today, which, as we have seen, is marginal at best, but in their refractive power to allow us to envision sustainable futures. This temporal dimension is hugely important as it affords the biofuel assemblage the possibility of constantly reframing itself, in terms of future benefits, future technologies and future possibilities and promises. Here, expert knowledge and economic imperatives generate new contexts for biofuels. To turn to Latour again, 'a technological project is not in a context; it gives itself a context, or sometimes does not give itself one' (ibid.: 133).

There is no conspiracy here; there is no purposeful generating of narratives to trick people into thinking that biofuels are the future. There is no unitary sense of purpose within the global assemblage. Rather, the assemblage represents the sum of interests, rationalities and economic imperatives that shape the socio-technical biofuels system. These interests and imperatives coalesce, clash and are continually reworked in response to change. They are bent by power, and that partially reflects the topographies of power that shape our world. The biofuels assemblage does not shape a bulletproof vision of its own reality, although it may seem to; instead it is contingent on the interpretive work of experts who seek to discern meaning from meaning by connecting practice to policy ideas (see Mosse 2005). Thus, we can witness a constant reworking of expertise, new technologies, and policy and economic imperatives and framings. It is in this context that biofuels become such a powerful idea that they can connect molecular genomics to climate-change science, and local agricultural practice to global value chains and commodity markets; it is through the articulation of these connections themselves that biofuels become so powerful and ubiquitous.

While biofuels are framed as a solution, they will not provide a multi-purpose salve to a world of inequality, finite resources and over-consumption. They simply cannot, for just as biofuels inhabit a world of ideas, we also inhabit a world of natural laws. Biofuels may, however, provide an insight into such a world. This book has attempted to articulate what some of these insights might be.

Science

The promises and risks of biofuels have exposed the limits of our knowledge. Untrammelled investment in new technology is driving new interactions with ecosystems, economies and collective and individual futures. We have invested not just in a technology, but also in our belief that it will allow us to deal with complex problems for which solutions otherwise look elusive. Biofuels offer a modern vision of how we deal with modern problems of our creation. This narrow perspective, of looking to first-, second- and third-generation technologies to deal with the world that confronts us, blinds us to the teleologies that led us there in the first place. We need to be clear that it is in understanding what led us there – whether we are thinking of climate change, underdevelopment or over-consumption – that we will ultimately find the solution. While there are undoubtedly proximate factors that shape unsustainability, there are also deeply rooted, historical structural factors with which we need to engage. We should not simply cover these with a veneer of science.

Looking beyond the present towards the future of next-generation technologies also blinds us to what we do not yet know. We apparently do not need to know or worry about the limits or deficiencies of current technologies because there will be new, better technologies tomorrow. Indeed, we need deficient current technologies if we are to unlock the promise of new generations of technologies. We need to be able to learn and to experiment and to make mistakes, although these activities generate contemporary implications and future trajectories. It is quite clearly developing countries and their populations who are the main participants in this experimentation. The world should not become a planetary laboratory.

This perspective, of placing faith in future technologies, appears to parallel many of the issues that have led us to this point in the first place: overconfidence in modernity, over-consumption, and the lack of an appreciation of our own limits. The promise of technology lures us into looking for the solutions of tomorrow rather than making difficult decisions today. Next-generation biofuels, which may be built on next-generation biotechnologies or synthetic biology, will allow us to continue to live the lives we choose, in the wealthy parts of the world at least. Consequences become inconsequential in the face of the promise of modernity.

We need to think far more carefully about the science. We need to invest in learning, knowledge and innovation, not in massive subsidies or policy prescriptions that drive forward imperfect, inappropriate or unsustainable science. Thinking more carefully and moving more purposefully will be far more effective in the longer term than the current situation, where progress is driven by subsidies – many of which were not even intended to stimulate the activities they are stimulating – investment opportunities and maintenance of a status quo that is clearly unsustainable. We need to strip science out of the future and understand that investing in it now is a way to shape our future, not bend to it. Science and technology are enormously powerful, and therein lie the risk and the promise. We need to find far more effective ways to capture that power and potential, not perpetually chase after it.

Systems

Subsidies, policy targets, investments and processing plants are accelerating biofuel development far ahead of our ability to comprehend the implications of what we are doing. We are no more able to be certain of whether US corn bioethanol, for example, is actually better or worse for the environment than fossil fuels than we are of being able to develop more appropriate technologies. Indeed, until we can sure what 'appropriate' or 'inappropriate' might actually mean in multiple contexts, we cannot be sure of very much at all.

It is clear that biofuels pose a range of complex systemic implications. They engage highly sensitive, highly complex systems and the resultant feedback loops, checks and balances are thus far too little understood. When one considers the complexities of indirect land-use change, how farmers elsewhere might react to changes in commodity prices, what this means for carbon emissions, and what the implications might be for food prices and therefore food security, it is easy to see the difficulties in identifying potential impacts, let alone modelling them.

It is, however, very difficult to create policy, plan and prioritize without data or analysis. Pathbreaking technologies entail profound implications and require appropriate management. Current methods of analysing the life-cycles of biofuels tend to be linear and limited to comparative analyses with fossil fuels. We need more nuanced methodologies that capture the complexities of broader interactions and implications. We also, frankly, need analyses that are independent from the interests that steer systems towards particular destinations.

Attempts are being made to understand complexity, but one cannot help but feel we should not have travelled so far down one path before working out that our means of navigation are not properly calibrated. It is difficult, if not impossible, to pull back, so we risk being stuck in an analytical tautology where we are continually trying to understand what is yet to happen in retrospect, after we have invested, planted and subsidized. We need to move towards a more reasoned, planned approach.

Historically, successful technologies have succeeded because they have been prompted and nurtured towards ubiquity or purpose (Latour 1996). Biofuels, with their complex assemblages of policymakers, entrepreneurs, experts and researchers supporting them, present an example of this. Perhaps a technology with as many implications as biofuels needs a more organized approach to its governance? We need to think of mechanisms through which biofuel development can be prioritized towards common needs, and not just driven by narrow interests. We need to develop ways to build circumspection into systems.

Synergy

Biofuels interact with multiple systems and are sustained through multiple systems. Part of the appeal of biofuels is their broad applicability and seeming ability to sidestep contradictions – they contribute to environmental care but drive the economy, they allow us to drive large cars in the USA and build sustainable livelihoods in Africa, they allow us to plan for the future by not changing very much today. Biofuels are able to tap into multiple, and what in the past would have been competing, networks of interests in a way that other technologies have not been able to do.

These new synergistic networks create new relationships between hitherto disconnected communities and interest groups. Thus, we find groups of farmers in Indonesia clearing trees to harvest palm oil to fuel European cars and salve European environmental consciences, or we witness 200 million more people going hungry in 2009/10 because of an unforeseen interaction between fixed US bioethanol subsidies and rising crude oil prices (at least to some extent).

Synergies create inequities too. Risk and responsibility flow through new networks like capital, or knowledge, or memes. Tanzanian livelihoods and access to land are potentially altered for ever to allow patterns of consumption and driving to work in private vehicles to *not* alter in Europe. Climate change has been driven by the cumulative effect of the historical release of GHGs, mainly in western Europe and North America; the social, economic and environmental implications of this are far greater for poorer countries and communities. Synergies then create networks in which these communities are obliged or at least encouraged to accept new risks (such as farming new commodities) to avoid other risks for which they should not feel any direct responsibility. Responsibility becomes a new global market and biofuels are the currency; responsibility can be absolved by purchasing biofuels and responsibility is taken by producing feedstocks. Unfortunately, this taking and giving of responsibility does not correlate with the relative production of GHGs, nor does it help those most at risk from the impacts of climate change.

This rewriting of a global script of responsibility creates inherent unsustainabilities in our ability to deal with the future. We cannot legitimately expect the poorest members of global society, those who are generally most risk averse and vulnerable to change, to accumulate risk and take responsibility for everyone. In doing so we are not allocating responsibility to those who have been most responsible or who have the greatest capacity to deal with the consequences of climate change. Instead, we are amplifying the risk to those who are already most vulnerable to these consequences, all in the name of a sustainability that is unsustainable, built as it is on a technology that is not yet proven, and the consequences of which are little understood. There are moral issues at stake here that should not be taken lightly, but beyond that we are also reducing our own collective ability to deal with the future.

These new networks stitch the world together in profound new ways. Since the onset of the era of globalization we have often talked of new ways in which the world comes together, and new responses and resistances to these processes. The global development of biofuels represents possibly the most profound reforging of North–South relationships since the height of colonialism. It is somewhat trite to say that new relationships that mirror colonial relationships are in many respects built on those legacies of industrialization, extraction and coercion. It is instructive, however, to reflect on those legacies and what they mean. In some respects, the discourses of sustainability that drive biofuels are built on them, unsustainable or inequitable as they may be.

Scale

These synergies do more than connect interests and actors in new ways; they also connect and reframe different scales. The globalization of concern about climate change and the environment has ignited interest in exploring global solutions, if not the political will to do so. This is one of the dynamics that biofuels are able to tap into. Thus, global attention has focused on global perspectives of biofuel development, with attendant concern about their implications. The potential of biofuels to do something other than just blend petroleum is almost invisible to us. Biofuels can play a more appropriate, sustainable role, providing energy to rural communities that currently have none. Witness rural Mali, or the potential of energy provision to communities who are not attached to electricity grids. We need to think beyond cars, especially private cars, and transportation. We need to think about using biofuels to generate power for those other than the powerful.

Similarly, some of the possible implications of investments in biofuels – 1 billion people being short of food, for example – are apparent only now the scale of the risk of food insecurity is perceived to be global. Food security is taken more seriously now we have recognized that rich countries may not simply be able to purchase their way around food shortages in future. The broader the array of networks that biofuels interact with, the more opportunities there are for new connections between the global and local to emerge, and scale therefore becomes an articulation of these new relationships.

Scale relates to broader issues about the way in which we should access energy with the environment in mind. Alternative, new sources of energy tend not to scale well, and simply cannot provide the power of fossil fuels without creating environmental implications or other negative consequences of their own. Consider the prospect of thousands of wind turbines within eyeline and earshot as opposed to one or two hidden over the horizon. Biofuels are a case in point; they simply do not – given present technologies – possess the potential to provide more than a small fraction of our energy needs, but they have come to be perceived as a 'global' energy source and a concern. All the biomass in the United States, *all* of it, can only ever meet a fraction of current energy demands. This translation of scale, from reality to theory, or potential to practical, presents problems in terms of how we manage sustainable development interventions in the future. Sustainability becomes altogether more problematic once we realize there are no simple solutions.

Finally, scale represents something much more profound in terms of our relationship to the environment via development. We simply have to collectively and individually acknowledge that we cannot continue to be so avaricious. Our pursuit of income, growth, resources and prosperity cannot be limitless. If it has taken the combination of climate change and a global financial crisis to realize this, it may be no bad thing; perhaps the realization of the widening gap between the promise and the reality of biofuels will be the prompt we need to move beyond realization to reality. Biofuels, despite their promise, will never scale up enough to allow us to live as we currently live. The laws of thermodynamics dictate this. And that, too, might be no bad thing, in retrospect.

Sustainability?

Asking whether biofuels are sustainable is a far less important question than asking whether we want them to be sustainable, or whether we are truly willing to try to lead sustainable lives and develop sustainable societies. We need to give things up; giving things up is the only way to move towards a sustainable future. We can hide behind stands of jatropha or sniff the promise of bioethanol, but neither will allow us to hide from the truth of the limits to growth in the longer term. Hiding behind the comfort of the compound of bio and fuels in the short term will solve nothing, and ultimately absolve no one.

Biofuels, our development of them, promotion of them and contestations over them, reflect nothing more than our vision of how we choose to relate to the world and how we choose to develop; not in a pre-planned, 'global' sense, but in an organic, fluid sense that reflects our priorities and the means and needs we choose to prioritize.

One cannot help feeling that we risk going round in circles. Our promotion and subsequent negation of biofuels reflect many earlier failed initiatives and unrequited promises. Our hope in tomorrow will, by definition, be forever self-unfulfilling. Our hopes for development, justice and environmental care will never be enough in themselves to overcome the realities and choices that would be required to make them so.

Biofuels demonstrate our unparalleled and growing ability to reshape the world, our relationship to it, and to each other. As global challenges deepen, our potential and resolve to create transformative technologies to deal with them strengthen, but we should remember that that, in itself, does not mean that we *can* deal with them. Unfortunately it is not so clear that our resolve to make the necessary decisions and compromises will strengthen, and while it does not we will continue to be drawn to the allure of biofuels and other perceived techno-fixes. We have to recognize that we need to make choices that will have a negative impact on the lives of those who consume the most. That is the unpalatable truth we hope to partially obscure through the consumption of biofuels.

I co-organized an informal workshop on bioenergy last year

at which a highly experienced engineer presented what he termed 'back of the cigarette packet' calculations of the global potential of biofuels (which almost perfectly tallied with similar data presented on biofuels' potential contribution to the UK's energy demands presented in Chapter 1), and talked about his frustration at the profoundly unintuitive notion, from a technical perspective, of burning biomass to convert into liquid in order for it to be combusted. He rounded off his talk by asking 'Why do policy-makers never listen to engineers?' The answer, of course, is that policy, and human nature, is driven by exigencies other than the First or Second Law of Thermodynamics. That is what we must overcome.

The globalization of risk

Biofuels risk generating not energy but a false sense of sustainability. Biofuels promise a supposedly radical new way of generating energy in the most benign manner possible, by changing very little at all. They allow us to continue on current pathways, accentuating current relationships and dependencies. We can over-consume but blend our way out of it. We can expend too much energy, but trade our way out of it. We can generate new environmental externalities, but innovate our way around them. We can ignore the most pressing of responsibilities, but identify others who can deal with them. Poverty, hunger and vulnerability remain hidden in the world. They are unseen and most of the time not thought about or recognized (Sachs 1999). Sustainability, too, remains an invisible, vague, woolly concept. Fewer and fewer people believe in the threat or even the reality of climate change. Energy, and petrol, is amorphous. For most of us, petrol goes from pump to tank without a moment's thought. If we pause for a moment and connect these things up it can be instructive. In reality these things *are* connected, and profoundly so. Biofuels strengthen these connections and generate new connections. Investment in biofuels risks amplifying the contours of poverty and consumption that shape the world, at the same time as they risk blinding us to these topographies.

Biofuels threaten an impenetrable triple-bind for developing countries and their citizens. First, the failure of biofuels to temper GHG emissions will pose the gravest threat to people in developing countries as they will suffer the greatest adverse effects of continued global warming. Second, biofuel development rests on the logics and extension of existing global agro-food chains that accumulate risk in the act of production and profit in the act of processing and transportation. Third, biofuels policy places responsibility for controlling GHG emissions with those who have not been and still are not responsible for producing an equal share of them. This bind of risk, responsibility and impact is entirely inequitable and cannot be sustainable or just. It echoes the past, of colonialism's creation of new relationships and generation of inequalities, and underlines the present, where overdevelopment and underdevelopment exist hand in hand.

We need to map the topographies that shape our relation to the world, and explore our connections with each other. Biofuels, in their current incarnation at least, do not promise sustainability, but perhaps they can expose us to the realities that confront us, and decisions that need to be made. The risks biofuels pose are multi-dimensional; they threaten the environment, equity, development, responsibility and ultimately society. Sustainability, if it is to mean anything at all, must strengthen those very dimensions. Biofuels are not a key to this, but are perhaps a keyhole through which we can peer directly at the problem, possibly for the first time. We need to establish a line of sight.

This book began by referring to the notion of a 'perfect storm', curiously bereft of human influence. It concludes by paraphrasing a line by Michael Watts in describing Mike Davis's *Late Victorian Holocausts*: 'Rosa Luxemburg meets the perfect storm' (Watts 2001). We need to focus a bit less on nature and a bit more on society and the relations of power that structure it if we are to derive sustainability from science, and equity over time.

Notes

1 Introduction

1 Associated Press, 27 October 2007.

2 *Guardian*, 24 January 2010.

3 Tropical biomass is on average five times more productive than temperate biomass (see Johnson and Yamba 2005).

4 *New York Times*, 30 May 2008.

5 Defined as the moment at which peak oil extraction is reached, indicating a future terminal decline in production.

2 Science

1 'Brazil disputes cost of sugar in the tank', *Guardian*, 10 June 2008.

2 Cited in G. Monbiot, 'A lethal solution', *Guardian*, 27 March 2007.

3 Earth Policy Institute, 'Data highlights: U.S. feed one quarter of its grain to cars while hunger is on the rise', www.earthpolicyinstitute. org.

4 Tanzania Investment Centre.

5 Quoted in Green, Inc., 'Tanzania suspends biofuels invest-ments', 14 October 2009.

6 'The next great land sale', *Africa-Asia Confidential*, 2(12), 26 January 2010, p. 6.

7 'Call for delay to biofuels policy', *BBC News*, 24 March 2008.

4 Synergy

1 EU Directive 2003/30 (2008).

2 EPA website, www.epa.gov/ OMS/renewablefuels/.

3 'A milestone on the road to green fuel', *Independent*, 27 June 2007.

4 FAO newsroom, Rome, 25 April 2006.

5 US Energy Information Administration and Bureau of Labor Statistics (data derived from Brent spot statistics).

6 Food, Conservation, and Energy Act of 2008 (Pub.L. 110-234, 122 Stat., 923, enacted 22 May 2008, H.R. 2419).

7 'EU rejects UK call for mora-torium on biofuels', *ICIS News*, 21 January 2008.

8 Associated Press, 27 October 2007.

9 The Center for Responsive Politics (n.d.), cited in Mol (2007).

10 J. Weber, 'The downside of ADM's focus on biofuels', *Business Week*, December 2008.

11 D. Childs, 'Chevron pumps more money into university biofuel research', *Cleantech*, May 2007.

12 Ibid.

13 'Shell announces six new biofuels research agreements', Shell press release, 17 September 2008.

14 See Luxemburg (1913) and Polanyi (1944), for example.

15 United States International Trade Commission data.

16 Although it is not altogether

clear how long these levels of proposed subsidy can or will be maintained.

17 'Deals can be good news when not made behind closed doors', *Guardian*, 7 March 2010.

18 Given the issues in data collection outlined above these should be considered conservative estimates.

19 'How food and water are driving a 21st century African land grab', *Observer*, 7 March 2010.

20 'Sinopec reportedly to invest US$5 billion in biofuels in Indonesia', *Biopact*, 19 October 2008.

21 See 'Seoul search in Africa', *Africa-Asia Confidential*, 1(13): 2; 'Madagascar leader axes land deal', BBC, 19 March 2009; J. Blas, 'Land leased to secure crops for South Korea', *Financial Times*, 18 November 2008.

22 T. Burgis, 'Lonrho secures rice land deal in Angola', *Financial Times*, 16 January 2009.

23 Reuters, 'Saudi firm in $400m farm investment in Africa', 15 April 2009.

24 Reuters, 'Saudi Hail starts farm investment abroad in Sudan', 16 February 2008.

25 Reuters, 'Firm eyes sorghum ethanol plant', 19 September 2008.

26 'Long road ahead before African biofuels take off', www.pangealink.org/african-biofuel-potential.html, accessed 8 March 2010.

27 'Emami Biotech to set up biofuel project in Ethiopia', *Business Standard*, Kolkata, 4 August 2009.

28 'Africa–India biofuel initiatives', Presentation prepared for Chatham House, April 2010.

29 'Eight hundred million euros for modular bio-power in Mozambique', *Biotech News*, 7 December 2006.

30 'How food and water are driving a 21st century African land grab', *Observer*, 7 March 2010.

31 University of Leeds and the UK Motor Industry Research Association, 'External vehicle speed control', 2000, cited in European Federation for Transport and Environment, 'Road transport, speed and climate change', 2005.

5 Scale

1 This is particularly true of the landless or people with limited tenure, as they are more likely to have no other productive activities to fall back on.

2 For example, oilseed crops tend to employ relatively large numbers of people as they are less mechanized than other crops. Tree crops, by contrast, require much less labour than agricultural crops.

3 Reuters, 'Indonesia sees 2008 fuel subsidy bill rising', 18 February 2008.

4 Increases in the areas of oil palm planted are essential if the EU is to meet its 10 per cent blending target as it is not possible for per-hectare yields to be increased sufficiently to achieve this (contrary to statements by the EU Commission).

5 It is very unlikely that the irregular environmental and certification practices identified in Indonesia would prevent the EU from importing as EU biofuel policies do not currently legislate against palm oil from plantations

developed on protected forest lands (FoE Netherlands 2009).

6 'Oil boom threatens the last orangutans', *Independent*, 23 June 2009.

7 K. Bradsher, 'A new, global oil quandary: costly fuel means costly calories', *New York Times,* 19 January 2008.

8 R. Mahabir, 'Failed policies knock biodiesel production by 85 per cent', *Jakarta Post*, 24 January 2008.

9 Bradsher, 'A new, global oil quandary'.

10 'Mali-Folk Centre converts pick-up to run on plant oil', African Centre for Plant Oil Technology, Nordic Folkcentre for Renewable Energy, Hurup Ty, Denmark, 15 November 2001.

11 L. Polgreen, 'Mali's farmers discover a weed's potential power', *New York Times*, 9 September 2007.

12 Out of fifty-four Poverty Reduction Strategy Papers reviewed in 2006, only that dealing with Mali makes reference to poverty reduction in relation to energy provision (UNDP 2006).

13 Reuters, 'Kenya court halts $370mn sugar, biofuels project', 13 July 2008.

14 Personal communication, James Pattison, PhD researcher, University of Edinburgh, March 2010.

15 I. Sachs, quoted in 'An in-depth look at Brazil's social seal fuel', 2007, biopact.com/2007/03/in-depth-look-at-brazils-social-fuel.htm.

6 Sustainability?

1 A. Jha, 'Gene scientist to create algae biofuel with Exxon Mobil', *Guardian*, 14 July 2009.

2 N. Wade, 'Researchers say they created a synthetic cell', *New York Times*, 20 May 2010.

Bibliography

ActionAid (2008) *Cereal Offenders? G8 Leaders on Biofuel/Hunger Charges*, London: ActionAid.

— (2010) *Meals per Gallon: The Impact of Industrial Biofuels on People and Global Hunger*, London: ActionAid.

Adams, J., C. King and N. Ma (2009) 'China: research and collaboration in the new geography of science', Leeds: Thomson Reuters.

Aldous, P. (2007) 'Interview: DNA's messengers', *New Scientist*, 2626: 57.

Alston, J., S. Dehmer and P. Pardey (2006) 'International initiatives in agricultural R&D: the changing fortunes of the CGIAR', in P. Pardey, J. Aston and R. Piggot (eds), *Agricultural R&D in the Developing World: Too Little, Too Late?*, Washington, DC: IFPRI.

Altenbuerg, T. et al. (2009) *Biodiesel in India*, German Development Institute, www.die-gdi.de.

Beck, U. (1992) *Risk Society: Towards a New Modernity*, London: Sage.

Beddington, J. (2009) 'Food security: a global challenge', Paper given at a BBSRC workshop on 'Food security', London, 19 February.

Bello, W. (2009) *The Food Wars*, London: Verso.

Bishop, M. and M. Green (2008) *Philanthrocapitalism: How the Rich Can Save the World and Why We Should Let Them*, London: A. & C. Black.

Cadenas, A. and S. Cabezudo (1998) 'Biofuels as sustainable technologies: perspectives for less developed countries', *Technological Forecasting and Social Change*, 58(1): 83–104.

Callon, M. (1986) 'Some elements of a sociology of translation: domestification of the scallops and fishermen of St Brieuc Bay', in J. Law (ed.), *Power, Action and Belief: A New Sociology of Knowledge?*, Sociological Review Monograph, Keele: University of Keele, pp. 196–233.

Castells, M. (1996) *The Rise of the Network Society*, London: Blackwell.

CFC (2007) *Biofuels: Strategic Choices for Commodity Dependent Developing Countries*, Amsterdam: Common Fund for Commodities.

CGIAR (2008) *Biofuels Research in the CGIAR: A Science Council Perspective*, Rome: Science Council of the Consultative Group on International Agricultural Research.

Christian Aid (2009) *Growing Pains: The Possibilities and Problems of Biofuels*, London: Christian Aid.

Clancy, J. (2008) 'Are biofuels pro-poor? Assessing the evidence', *European Journal of Development Research*, 20(30): 416–31.

CONCAWE (Oil Companies'

European Association for Environment, Health and Safety in Refining and Distribution, but the acronym is derived from 'Conservation of Clean Air and Water in Europe'), Joint Research Centre of the EU Commission, and European Council for Automotive R&D (2004) *Well-to-Wheels Analysis of Future Automotive Fuels and Powertrains in the European Context*, Version 1b, January, ies.jrc.cec.eu.int/Download/eh.

Conceição, P. and R. Mendoza (2009) 'Anatomy of the global food crisis', *Third World Quarterly*, 30(6): 1159–82.

Cotula, L., N. Dyer and S. Vermeulen (2008) *Fuelling Exclusion? The Biofuels Boom and Poor People's Access to Land*, London: IIED.

Cotula, L., S. Vermeulen, R. Leonard and J. Keeley (2009) *Land Grab or Development Opportunity? Agricultural Investment and International Arms Deals in Africa*, London: IIED.

Coyle, W. (2007) 'The future of biofuels: a global perspective', *Amber Waves*, US Department of Agriculture.

Crutzen, P., A. Mosier, K. Smith and W. Winiwarter (2008) 'N_2O release from agro-biofuel production negates global warming reduction by replacing fossil fuels', *Atmospheric Chemistry and Physics*, 8(2): 389–95.

Dauvergne, P. and K. Neville (2009) 'The changing pattern of north–south and south–south political economy of biofuels', *Third World Quarterly*, 30(6): 1087–102.

De Fraiture, C., M. Giordano and

L. Yongsong (2008) 'Biofuels: implications for agricultural waste water use: blue impacts of green energy', *Water Policy Supplement*, 1: 67–81.

De La Torre Ugarte, D. (2006) 'Developing bioenergy economic and social issues: bioenergy and agriculture promises and challenges', 2020 *Vision Briefs*, 14(2), Washington, DC: International Food Policy Research Institute.

Delucchi, M. A. (2003) *A Lifecycle Emissions Model (LEM): Lifecycle Emissions from Transportation Fuels, Motor Vehicles, Transportation Modes, Electricity Use, Heating and Cooking Fuels, and Materials – Documentation of Methods and Data*, UCD-ITS-RR-03-17, Davis: Institute of Transportation Studies, University of California.

Demirbas, A. (2007) 'Progress and recent trends in biofuels', *Progress in Energy and Combustion Science*, 33: 1–18.

Dufey, A. (2007) *International Trade in Biofuels: Good for Development? And Good for Environment?*, IIED Briefing, London.

Escobar, A. (2009) *Territories of Difference: Place, Movements, Life*, Durham, NC: Duke University Press.

European Union (2007) *Communication from the Commission to the Council and the European Parliament/Renewable Energy Roadmap/Renewable Energies in the 21st Century: Building a More Sustainable Future*, 10 January.

FAO (2000) 'The energy and agriculture nexus', Environment

and Natural Resources Working
Paper no. 4, Rome: FAO.
— (2009a) *The State of Food
Insecurity in the World 2009*,
Rome: FAO.
— (2009b) *Small Scale Bioenergy
Initiatives: Brief Description and
Preliminary Lessons on Liveli-
hood Impacts from Case Studies
in Asia, Latin America and
Africa*, Rome: FAO/PAC/PISCES.
Fargione, J., J. Hill, D. Tilman,
S. Polasky and P. Hawthorne
(2008) 'Land clearing and the
carbon debt', *Science*, 319(5867):
1235–8.
Farrell, A., R. Plevin, B. Turner,
B. Jones, M. O'Hare and
D. Kammen (2006) 'Ethanol
can contribute to energy and
environmental goals', *Science*,
311: 506–8.
FoE (2005) *The Oil for Ape Scandal:
How Palm Oil is Threatening the
Orang-utan*, London: Friends of
the Earth.
FoE Netherlands (2009) *Failing
Governance – Avoiding Res-
ponsibilities: European Biofuel
Policies and Oil Palm Plantation
Expansion in Ketapang District,
West Kalimantan*, Amsterdam:
Friends of the Earth Netherlands
and WALHI KalBar.
Frow, E., D. Ingram, W. Powell, D.
Steer, J. Vogel and S. Yearley
(2009) 'The politics of plants',
Food Security, 1(1): 17–23.
Fulton, L. et al. (2004) *Biofuels for
Transport: An International
Perspective*, Paris: International
Energy Agency.
Gerbens-Leenes, P., A. Hoekstra
and T. van der Meer (2009)
'Water footprint of bioenergy
and other primary energy car-
riers', Value of Water Research

Report series, no. 29, Delft:
UNESCO.
Giampietro, M. and K. Mayumi
(2009) *The Biofuel Delusion:
The Fallacy of Large-scale Agro-
Biofuel Production*, London:
Earthscan.
Gibson, D., D. Glass, J. Venter et al.
(2010) 'Creation of a bacterial
cell controlled by a chemically
synthesized genome', *Science*
online, 20 May.
Giddens, A. (2009) *The Politics of
Climate Change*, London: Polity
Press.
Giovannucci, D. and S. Ponte (2005)
'Standards as a new form of
social contract? Sustainability
initiatives in the coffee industry',
Food Policy, 30(3): 284–301.
Goldemberg, J. (2006) 'The ethanol
program in Brazil', *Environmen-
tal Research Letters*, 1.
Goldemberg, J., S. Coehlo,
P. Nastari and O. Lucon (2003)
'Ethanol learning curve – the
Brazilian experience', *Biomass
Bioenergy*, 26(3): 301–4.
Gonsalves, J. (2006) 'An assessment
of the biofuels industry in
India', United Nations Confer-
ence on Trade and Development,
18 October.
Gordon-Maclean, A., J. Laizer,
P. Harrison and R. Shemdoe
(2008) *Biofuel Industry Study,
Tanzania*, Tanzania and Sweden:
World Wide Fund for Nature
(WWF).
Green, R., S. Cornell, J. Scharle-
mann and A. Balmford (2005)
'Farming and the fate of wild
nature', *Science*, 307: 550–55.
Greene, N. (principal author) (2004)
*Growing Energy: How Biofuels
Can Help End America's Oil
Dependence*, New York: Natural

Resources Defense Council, December.

Hagens, N., R. Costanza and K. Mulder (2006) 'Letter in response to Farrell, et al.', *Science*, 312: 1747.

Hamelinck, C. N. et al. (2005) 'Ethanol from lignocellulosic biomass: techno-economic performance in short-, middle- and long-term', *Biomass and Bioenergy*, 28: 384–410.

Harrar, J., P. Mangelsdorf and W. Weaver (1952) 'Notes on Indian agriculture', Royal Agricultural College Archive, 11 April.

HM Treasury (2008) *The King Review of Low Carbon Cars*, London: Government Printers.

Howarth, R. and S. Bringezu (2009) 'Biofuels and environmental impacts. Scientific analysis and implications for sustainability', Policy Brief Series, UNESCO-SCOPE-UNEP.

Howse, R., P. van Bork and C. Hebebrand (2006) 'WTO disciplines and biofuels: opportunities and constraints in the creation of a global marketplace', Washington, DC: International Food and Agricultural Trade Policy Institute.

Hulme, M. (2009) *Why We Disagree about Climate Change: Understanding Controversy, Inaction and Opportunity*, Cambridge: Cambridge University Press.

International Energy Agency (2006) *World Energy Outlook 2006*, Paris: International Energy Agency.

— (2009) *World Energy Outlook 2009*, Paris: International Energy Agency.

Jessop, B. (1998) 'The rise of governance and the risks of failure: the case of economic development', *International Social Sciences Journal*, 50(155): 30–45.

Johnson, F. and F. Yamba (2005) 'Comparative advantage in the production of biofuels', *Renewable Energy for Development*, Stockholm: SEI.

Jongschaap, R., P. Corré, P. Bindraban and W. Brandenburg (2007) 'Claims and facts on jatrophs curcas L.: global jatropha curcas evaluation, breeding and propagation programmes', *Plant Research International*, 158: 1–42.

Jordan, A., R. Wurzel and A. Zito (2005) 'The rise of "new" policy instruments in comparative perspective: has governance eclipsed government?', *Political Studies*, 53(3): 477–96.

Junginger, M. et al. (2008) 'Developments in international bioenergy trade', *Biomass and Bioenergy*, 32: 717–29.

Kadam, K. L. (2002) 'Environmental benefits on a life cycle basis of using bagasse-derived ethanol as a gasoline oxygenate in India', *Energy Policy*, 30(5): 371–84.

Kamanga, K. (2008) *The Agrofuel Industry in Tanzania: A Critical Enquiry into Challenges and Opportunities. A Research Report*, Dar es Salaam: Hakiardhi and Oxfam Livelihoods Initiative for Tanzania (JOLIT).

Kammen, D., R. Bailis and A. Herzog (2001) 'Clean energy for development and economic growth: biomass and other renewable energy options to meet energy and development needs in poor countries', UNDP

Policy Discussion Paper, New York.

Kartha, S., G. Leach and S. Rajan (2005) *Advancing Bioenergy for Sustainable Development: Guidelines for Policymakers and Investors*, Washington, DC: World Bank.

Kaufmann, R. (2006) 'Letter in response to Farrell, et al.', *Science*, 312: 1747.

Kehati Foundation (2007) *Revising the Hope: Review on Bio-fuel Development Policy and Its Role in Policy Reduction in Indonesia.*

Kill, J. (2007) 'Biofuels are not the answer', Transnational Institute website.

Kojima, M. and T. Johnson (2005) *Potential for Biofuels for Transport in Developing Countries*, Washington, DC: World Bank.

Koplow, D. (2007) 'Biofuels – at what cost? Government support for ethanol and biodiesel in the United States: 2007 update', Geneva: Global Subsidies Initiative of the International Institute for Sustainable Development.

Kovarik, B. (1998) 'Henry Ford, Charles F. Kettering and the fuel of the future', *Automotive History Review*, 32: 7–27, www.radford.edu/~wkovarik/papers/fuel.html.

Kumar Biswas, P., S. Pohit and R. Kumar (2010) 'Biodiesel from jatropha: can India meet the 20% blending target?' *Energy Policy*, 38(3): 1477–84.

Kutas, G., C. Lindberg and R. Steenblik (2007) 'Biofuels – at what cost? Government support for ethanol and biodiesel in the European Union', Geneva: Global Subsidies Initiative of the International Institute for Sustainable Development.

Larson, E. (2006) 'A review of life-cycle analysis studies on liquid biofuel systems for the transport sector', *Energy for Sustainable Development*, X(2).

Latour, B. (1992) 'Where are the missing masses? The sociology of a few mundane artefacts', in W. Bijker and J. Law (eds), *Shaping Technology/Building Society: Studies in Sociotechnical Change*, Cambridge, MA: MIT Press.

— (1996) *Aramis. Or the Love of Technology*, Cambridge, MA: Harvard University Press.

Leach, M. and I. Scoones (2006) *The Slow Race: Making Technology Work for the Poor*, London: Demos.

Luxemburg, R. (1913) *The Accumulation of Capital*, London.

Luxresearch (2010) 'Ranking biofuel startups on the Lux innovation grid', Lux Research Report.

Macedo, I. C., M. R. L. Leal and J. E. A. R. da Silva (2004) *Assessment of Greenhouse Gas Emissions in the Production and Use of Fuel Ethanol in Brazil*, São Paulo: São Paulo State Secretariat of the Environment, May.

MacKay, D. (2009) *Sustainable Energy – without the Hot Air*, Cambridge: UIT.

Mackenzie, D. (2009) 'Rich countries carry out "21st century land grab"', *New Scientist*, 2685: 8–9.

Mandal, R. (2005) *Energy – Alternate Solutions for India's Needs: Biodiesel*, New Delhi: Planning Commission, Government of India.

McCullough, E., P. Pingali and
  K. Stamoulis (eds) (2008) *The
  Transformation of Agri-Food
  Systems: Globalization, Sup-
  ply Chains and Smallholder
  Farmers*, London: Earthscan.
McMichael, P. (2009) 'The agrofuels
  project at large', *Critical Socio-
  logy*, 35(6): 825–39.
Meillassoux, C. (1981) *Maidens,
  Meals and Money: Capitalism
  and the Domestic Community*,
  Cambridge: Cambridge Univer-
  sity Press.
Melillo, J., A. Gurgel, D. Kick-
  lighter, J. Reilly, T. Cronin,
  B. Felzer, S. Paltsev, C. Schlosser,
  A. Sokolov and X. Wang (2009)
  'Unintended environmental
  consequences of a global
  biofuels programme', Report
  no. 168, Cambridge, MA: MIT
  Joint Programme on the Science
  and Policy of Global Change.
Menichetti, E. and M. Otto (2008)
  'Existing knowledge and limits
  of scientific assessment of the
  sustainability impacts due to
  biofuels by LCA methodo-
  logy', Final report for the
  United Nations Environment
  Programme.
Ministry of Energy (2008) *Bio-
  diesel Strategy for Kenya*, Draft,
  Nairobi: Kenyan Government.
Mitchell, D. (2008) 'A note on rising
  food prices', Washington, DC:
  World Bank.
Modi, V., S. McDade, D. Lallement
  and J. Saghir (2006) *Energy and
  the Millennium Development
  Goals*, Energy Sector Manage-
  ment Assistance Programme,
  United Nations Development
  Programme, UN Millennium
  Project and World Bank.
Mol, A. (2007) 'Boundless
  biofuels? Between environmental
  sustainability and vulnerability',
  Sociologia Ruralis, 47(4):
  297–315.
— (2010) 'Environmental authorities and biofuel controversies',
  Environmental Politics, 19(1):
  61–79.
Molony, T. and J. Smith (2010)
  'Biofuels, food security and
  Africa', *African Affairs*, 109(436):
  489–98.
Moore, D. (2008a) 'Biofuels are
  dead: long live biofuels(?) – part
  one', *New Biotechnlogy*, 25(1):
  6–12.
— (2008b) 'Biofuels are dead: long
  live biofuels(?) – part two', *New
  Biotechnology*, 25(2/3): 96–100.
Moreira, J. (2006) 'Brazil's experi-
  ence with bioenergy', Brief 8 in
  P. Hazell and R. Pachauri (eds),
  *Bioenergy and Agriculture:
  Promises and Challenges*, Wash-
  ington, DC: International Food
  Policy Research Institute.
Mosse, D. (2005) *Cultivating
  Development: An Ethnography
  of Aid and Practice*, London:
  Pluto Press.
Motaal, D. (2008) 'The biofuels
  landscape: is there a role for the
  WTO?', *Journal of World Trade*,
  42(1): 61–86.
Mouk, B., S. Kirui, D. Theuri and
  J. Wakhungu (2010) 'Policies
  and regulations affecting biofuel
  development in Kenya', PISCES
  Policy Brief no. 1, Nairobi.
Nitske, W. R. and C. M. Wilson
  (1965) *Rudolf Diesel, Pioneer
  of the Age of Power*, Norman:
  University of Oklahoma Press.
OECD/FAO (2008) *OECD-FAO
  Agricultural Outlook 2008–2017*,
  Paris/Rome.
OECD/IEA (2008) *Energy Techno-*

logy Perspectives. Scenarios and Strategies to 2050, Paris.

Ong, A. and S. Collier (2005) Global Assemblages: Technology, Politics and Ethics as Anthropological Problems, London: Blackwell.

Openshaw, K. (2000) 'A review of Jatropha curcas: an oil plant of unfulfilled promise', Biomass and Bioenergy, 19: 1–15.

Oxfam (2001) Rigged Rules and Double Standards – Trade, Globalisation and the Fight against Poverty, Oxford: Oxfam.

— (2007) Biofuelling Poverty: Why the EU Renewable Fuel Target May be Disastrous for Poor People, Oxford: Oxfam International.

— (2008) 'Another inconvenient truth: how biofuel policies are deepening poverty and accelerating climate change', Oxfam Briefing Paper 114, Oxford.

Padilla, A. (2007) Biofuels: A New Wave of Imperialist Plunder of Third World Resources, 5, Penang: People's Coalition on Food Sovereignty.

Patel, R. (2007) Stuffed and Starved: From Farm to Fork, the Hidden Battle for the World Food System, London: Portobello Books.

Perrow, C. (1999) Normal Accidents: Living with High-risk Technologies, Princeton, NJ: Princeton University Press.

Peskett, L., R. Slater, C. Stevens and A. Dufey (2007) 'Biofuels, agriculture and poverty reduction', Natural Resource Perspectives, 107, London: Overseas Development Institute.

Pimentel, D. (2004) 'Ethanol fuels: energy balance, economic and environmental impacts are negative', Natural Resources Research, 12(2): 127–34.

Pin Koh, L. and J. Ghazoul (2008) 'Biofuels, biodiversity and people: understanding the conflicts and finding opportunities', Biological Conservation, 141: 2450–60.

Polanyi, K. (1944) The Great Transformation, Boston, MA: Beacon Press.

Ponte, S. (2005) The Coffee Paradox: Global Markets, Commodity Trade and the Elusive Promise of Development, London: Zed Books.

Pousa, G., A. Santos and P. Saurez (2007) 'History and policy of biodiesel in Brazil', Energy Policy, 35: 5393–8.

Quirin, M., S. O. Gartner, M. Pehnt and G. A. Reinhardt (2004) 'CO$_2$ mitigation through biofuels in the transport sector: status and perspectives', Main report, Heidelberg: Institute for Energy and Environmental Research (IFEU).

Raguaskas, A., C. Williams, B. Davidson, G. Britovsek, J. Cairney, C. Eckert, W. Frederick, J. Hallett, D. Leak, C. Liotta, J. Milenz, R. Murphy, R. Templer and T. Tschplinski (2006) 'The path forward for biofuels and biomaterials', Science, 313(1742).

REN21 (2008) Renewables 2007 Global Status Report, Paris: REN21 Secretariat.

— (2009) Renewables 2009 Global Status Report, Paris: REN21 Secretariat.

Rothkopf, G. (2007) 'A blueprint for green energy in the Americas: strategic analyis of opportunities for Brazil and the hemisphere,

Global Biofuels Outlook, Washington, DC: Inter-American Development Bank.

Royal Society (2008) *Sustainable Biofuels: Prospects and Challenges*, Policy Document 01/08, London.

Sachs, J. (1999) 'Helping the world's poorest', *The Economist*, 14 August, pp. 17–20.

Searchinger, T., R. Heimlich, R. Houghton, F. Dong, A. Elobeid, J. Fabiosa, S. Tokgoz, D. Hayes and T. Yu (2008) 'Use of U.S. croplands for biofuels increases greenhouse gases through emissions from land-use change', *Science*, 319: 1238–340.

Searchinger, T., S. Hamburg, J. Melillo, W. Chameides, P. Havlik, D. Kamen, G. Likens, R. Lubrowski, M. Obersteiner, M. Oppenheimer, G. Robertson, W. Schlesinger and G. Tilman (2009) 'Fixing a critical climate accounting error', *Science*, 326: 527–8.

Sen, A. (1981) *Poverty and Famines: An Essay on Entitlement and Deprivation*, Oxford: Clarendon Press.

Shackleton, C., C. Shackleton, B. Buiten and N. Bird (2007) 'The importance of dry woodlands and forests in rural livelihoods and poverty alleviation in South Africa', *Forest Policy and Economics*, 9(5): 558–77.

Shapouri, H., J. Duffield and M. Wang (2002) 'The energy balance of corn ethanol: an update', Agricultural Economics Reports, Washington, DC: Office of the Chief Economist.

Shay, E. (1993) 'Diesel fuels from vegetable oils – status and opportunities', *Biomass and Bioenergy*, 4: 227–42.

Shiva, V. (2009) *Soil, Not Oil: Climate Change, Peak Oil and Food Insecurity*, London: Zed Books.

Shukla, S. (2008) 'Biofuel development programme of Chattisgarh', Fifth International Development Conference on Biofuels, Winrock International, India.

Smith, J. (2007) 'Culturing development: bananas, petri dishes and "mad science" in Kenya', *Journal of Eastern African Studies*, 1(2): 212–33.

— (2009) *Science and Technology for Development*, London: Zed Books.

— (2010) 'New institutional arrangements for development, science and technology', *Development*, 53(1): 48–53.

Smolker, R., B. Tokar and A. Petermann (2008) *The Real Cost of Agrofuels: Impacts on Food, Forests, Peoples and the Planet*, Global Forest Coalition and Global Justice Ecology Project.

Songela, F. and A. Maclean (2008) *Scoping Exercise on the Biofuels Industry within and outside Tanzania*, Energy for Sustainable Development Report for the WWF Tanzania Programme Office.

Steenblik, R. (2007) *Biofuels – At What Cost? Government Support for Ethanol and Biodiesel in Selected OECD Countries*, Global Studies Initiative of the International Institute for Sustainable Development, Geneva.

Stern, N. (2007) *The Economics of Climate Change: The Stern*

Review, Cambridge: Cambridge University Press.

Sticklen, M. (2006) 'Plant genetic engineering to improve biomass characteristics for biofuels', *Current Opinion in Biotechnology*, 17: 315–19.

Stoker, G. (1998) 'Governance as theory: five propositions', *International Social Science Journal*, 50: 17–28.

Stokstad, E. (2009) 'Agricultural science gets more money, new faces', *Science*, 326(5950): 216.

Sulle, E. and R. Nelson (2009) *Biofuels, Land Access and Rural Livelihoods in Tanzania*, London: IIED.

Tauli-Corpuz, V. and P. Tamang (2007) *Oil Palm and Other Commercial Tree Plantations, Monocropping: Impacts on Indigenous Peoples' Land Tenure and Resource Management Systems and Livelihoods*, United Nations Permanent Forum on Indigenous Issues, 6th Session, New York, 14–25 May.

TERI (2004) *Liquid Biofuels for Transportation: India Country Study on the Potential and Implications for Sustainable Agriculture and Energy*, New Delhi: Government of India.

— (2005) *Detailed Project Report for the National Mission on Biodiesel*, Prepared for the Department of Land Resources, Ministry of Rural Development, New Delhi: Government of India.

Tilman, D. et al. (2006) 'Biodiversity and ecosystem stability in a decade-long biodiversity experiment', *Nature*, 441: 629–32.

Tyner, W. (2007) 'Policy alternatives for the future biofuels industry',

Journal of Agricultural and Food Industrial Organization, 5(2).

— (2008) 'The US ethanol and biofuels boom: its origins, current status, and future prospects', *BioScience*, 58(7): 646–53.

Tyner, W. and J. Quear (2006) 'Comparison of a fixed and variable corn ethanol subsidy', *Choices*, 21: 199–202.

UNCTAD (2008) *World Investment Directory 2008*, vol. X: *Africa*, Geneva: United Nations Conference on Trade and Development.

UNDP (1995) *Energy as an Instrument for Socioeconomic Development*, New York: United Nations Development Programme.

— (2004) *Reducing Rural Poverty through Increased Access to Energy Services: A Review of Multifunctional Platforms in Mali*, New York: United Nations Development Programme.

— (2006) *Energizing Poverty Reduction: A Review of the Energy–Poverty Nexus in Poverty Reduction Strategy Papers*, New York: United Nations Development Programme.

UN-Energy (2007) *Sustainable Bioenergy: A Framework for Decision-makers*, New York: UN-Energy.

UNEP (2009) *Towards Sustainable Production and Use of Resources: Assessing Biofuels*, New York: United Nations Environment Programme.

UNEP and UNESCO (2007) *The Last Stand of the Orangutan. State of Emergency: Illegal Logging, Fire and Palm Oil in Indonesia's National Parks*, New York.

Van der Voet, E., R. Lifset and L. Luo (2010) 'Life-cycle assessment of biofuels, convergence and divergence', *Biofuels*, 1(3): 435–49.

Van Eijck, J. and H. Romijn (2008) 'Prospects of jatropha biofuels in Tanzania: an analysis with strategic niche management', *Energy Policy*, 36: 311–25.

Verdonk, M., C. Dieperink and A. Faaij (2007) 'Governance of the emerging bio-energy markets', *Energy Policy*, 35: 3909–24.

Vianello, M. (2009) 'Biofuels and development: exploring the Kenyan reality', Unpublished MSc dissertation, University of Edinburgh.

Wakker, E. (2004) *Greasy Palms: The Social and Ecological Impacts of Large Scale Oil Palm Development in Southeast Asia*, London: Friends of the Earth.

Watts, M. (2001) 'Black acts', *New Left Review*, 9: 125–40.

Wetlands International (2006) *Peatland Degradation Fuels Climate Change*, Wageningen: Wetlands Institute.

Wilkinson, J. and S. Herrera (2008) *Agrofuels in Brazil: What is the Outlook for its Farming Sector?*, Rio de Janeiro: CPDA/UFRRJ/Oxfam.

Wirl, F. (2009) 'OPEC as a political and economical entity', *European Journal of Political Economy*, 25(4): 399–408.

Wise, M., K. Calvin, A. Thomson, L. Clarke, B. Bond-Lamberty, R. Sands, S. Smith, A. Janetos and J. Edmonds (2009) 'Implications for limiting CO_2 concentrations for land use and energy', *Science*, 324: 1183–6.

Woods, J. and R. Diaz-Chavez (2007) 'The environmental certification of biofuels', Discussion Paper no. 2007-6, International Transport Forum, OECD.

World Bank (2010) *World Development Report 2010. Energy and Development*, Washington, DC: World Bank.

WorldWatch Institute (2007) *Biofuels for Transportation: Global Potential and Implications for Sustainable Agriculture and Energy in the 21st Century*, Washington, DC: WorldWatch Institute.

Yearley, S. (2005) *Making Sense of Science: Understanding the Social Study of Science*, London: Sage.

Index